Use of this workbook is limited to educational purposes.

This workbook may not be copied, sold, or resold from any person, entity, or organization other than the originating designer, John A. Honeycutt.

All content, images, and thought-leadership are owned by their respective copyright holder and subject to those rights in addition to restrictions imposed by John A. Honeycutt.

Processes associated with design and layout of curriculum is patent-pending by John A. Honeycutt as Honeycutt 21st Century Instructional Design. Design, layout, and content not otherwise owed by other entities is owned by John A. Honeycutt.

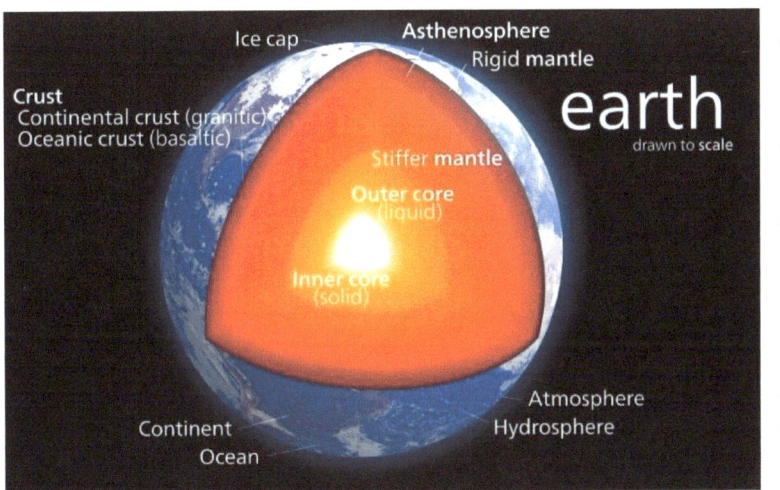

EARTH SCIENCE
HoneycuttScience Work Book

Copyright John A. Honeycutt 2017. All rights reserved.

Contents

11.1	What is Earth Science?	1
12.1	Scientific Method and Safety	7
13.1	Connections Across Content	13
14.1	Map Interpretation	19
16.1	Earth Chemistry	25
18.1	Rocks and Rock Types	31
19.1	Resources and Energy	37
21.1	The Rock Record	43
23.1	Plate Tectonics and Today's Earth	49
24.1	Deformation of the Crust	55
25.1	Earthquakes, Volcanoes & Tsunamis	61
26.1	Weathering, Erosion & Rivers	67
27.1	Agricultural Resources	73
28.1	Hydrocarbons and Energy	79

Strategic Thinking Pages	85

11.1 What is Earth Science?

Summarize main points from each video.

Video Title / topic

Video Title / topic

Video Title / topic

Topic Introduction

Summarize your understanding of each paragraph.

Earth is the mighty planet upon which we all live. Only recently have humans begun to understand the complexity of this planet. In fact, it was only a few hundred years ago that we discovered that Earth was just a tiny part of an enormous galaxy.

[ ]

Earth Science deals with any and all aspects of the Earth. Our Earth has molten lava, icy mountain peaks, steep canyons and towering waterfalls. Earth scientists study the atmosphere high above us as well as the planet's core far beneath us.

[ ]

Earth scientists study parts of the Earth as big as continents and as small as the tiniest atom. In all its wonder, Earth scientists seek to understand the beautiful sphere on which we thrive.

[ ]

Earth scientists specialize in studying just a small aspect of our Earth. Since all of the branches are connected together, specialists work together to answer complicated questions.

[ ]

https://en.wikibooks.org/wiki/High_School_Earth_Science/Earth_Science_and_Its_Branches

Read/Summarize Text

1. **Read the passage.**
2. **Underline key expressions in each sentence.**
3. **Re-write each word (or expression) you underlined.**
4. **Summarize the passage.**

Title of Passage.

> Geology is the study of the solid matter that makes up Earth. Anything that is solid, like rocks, minerals, mountains, and canyons is part of geology.
>
> Geologists study the way that these objects formed, their composition, how they interact with one another, how they erode, and how humans can use them. Geology has so many branches that most geologists become specialists in one area. For example, a mineralogist studies the composition and structure of minerals such as halite (rock salt), quartz, calcite, and magnetite .

https://en.wikibooks.org/wiki/High_School_Earth_Science/Earth_Science_and_Its_Branches

Re-write words you underlined

_____   _____   _____

_____   _____   _____

Using a complete sentence, summarize or rephrase the passage

Read Text for Comprehension

Read this article for deeper understanding. No summary is required, although you may want to circle, underline, or mark key ideas and words.

Oceanography is the study of everything in the ocean environment. More than 70% of the Earth's surface is covered with water. Most of that water is found in the oceans. Recent technology has allowed us to go to the deepest parts of the ocean, yet much of the ocean remains truly unexplored.

Climatologists help us understand the climate and how it will change in the future in response to global warming. Oceanographers study the vast seas and help us to understand all that happens in the water world. As with geology, there are many branches of oceanography. Physical oceanography is the study of the processes in the ocean itself, like waves and ocean currents

Meteorologists don't study meteors — they study the atmosphere! Perhaps this branch of Earth Science is strangely named but it is very important to living creatures like humans. Meteorology includes the study of weather patterns, clouds, hurricanes, and tornadoes. Using modern technology like radars and satellites, meteorologists work to predict or forecast the weather. Because of more accurate forecasting techniques, meteorologists can help us to prepare for major storms, as well as help us know when we should go on picnics.

Atmospheric scientists study the whole atmosphere, which is a thin layer of gas that surrounds the Earth. Most of it is within about 10 - 11 kilometers of the Earth's surface. Earth's atmosphere is denser than Mars's thin atmosphere, where the average temperature is -63° C, and not as thick as the dense atmosphere on Venus, where carbon dioxide in the atmosphere makes it hot and sulfuric acid rains in the upper atmosphere. The atmosphere on Earth is just dense enough to even out differences in temperature from the equator to the poles, and contains enough oxygen for animals to breathe.

Astronomers have proven that our Earth and solar system are not the only set of planets in the universe. By 2007, over a hundred planets outside our solar system had been discovered. Although no one can be sure how many there are, astronomers estimate that there are billions of other planets. In addition, the universe contains black holes, other galaxies, asteroids, comets, and nebula. As big as Earth seems to us, the entire universe is vastly greater. Our Earth is an infinitesimally small part of our universe.

https://en.wikibooks.org/wiki/High_School_Earth_Science/Earth_Science_and_Its_Branches

Draw Illustration

Copy and Label the Illustration in the Space Provided

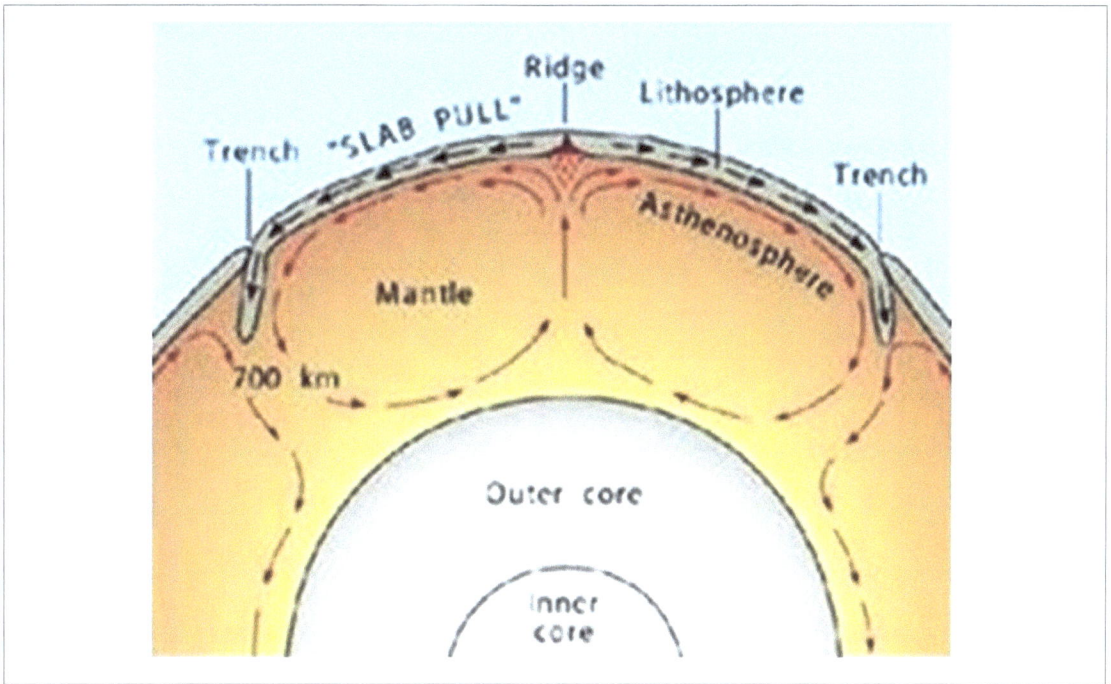

https://www.pinterest.com/pin/561683384748845787/

Draw (Copy) the Illustration Here

Interpret a Graph

Write the title of the graph _____

Circle the type of chart this represents

     Bar Chart   Line Chart   Pie Chart   Other

If applicable,

    What does the X-axis represent _____

    What does the Y-axis imply _____

Summarize what this graph represents or conveys

_____

http://www.rssweather.com

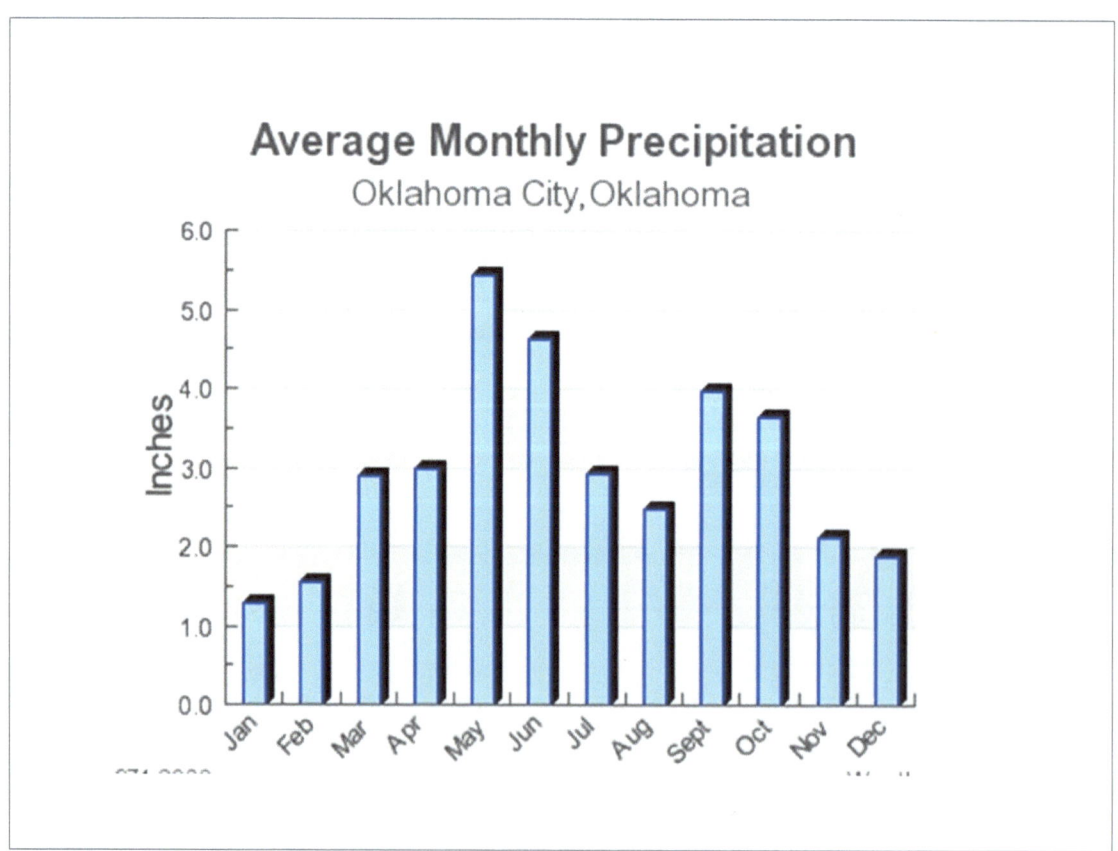

12.1 Scientific Method and Safety

Summarize main points from each video.

Video Title / topic

Video Title / topic

Video Title / topic

Topic Introduction

Summarize your understanding of each paragraph.

The scientific method consists of (1) making observations (2) writing down a hypothesis and (3) testing the hypothesis. When new observations do not support the original hypothesis, a new hypothesis is required.

[ ]

Sometimes, hypothesis are formulated before observations are collected; sometimes observations are made before hypothesis are created. Either way, it is important that scientists carefully record their procedures.

[ ]

In Earth Science classes in high school, student observations are sometimes made outside – or "in the field." Some science tests are conducted inside – in a lab environment. In both of these situations, safe practices are required.

[ ]

Identification of hazards and safety risks prior to beginning lab work, or making observations in the field is an important step. The best way to reduce risk is to eliminate them. But at a minimum, personal protective equipment must be work and following procedures.

[ ]

Read/Summarize Text

1. **Read the passage.**
2. **Underline key expressions in each sentence.**
3. **Re-write each word (or expression) you underlined.**
4. **Summarize the passage.**

About scientific hypothesis, observations, theories, and laws.

The scientific method is employed by scientists around the world, but it is not always conducted in the order above. Sometimes, hypothesis are formulated before observations are collected; sometimes observations are made before hypothesis are created. Regardless, it is important that scientists record their procedures carefully, allowing others to reproduce and verify the experimental data and results. After many experiments provide results supporting a hypothesis, the hypothesis becomes a theory. Theories remain theories forever, and are constantly being retested with every experiment and observation.

NOTE: Theories can never become fact or law.

cK12.org

Re-write words you underlined

_____   _____   _____

_____   _____   _____

Using a complete sentence, summarize or rephrase the passage

Read Text for Comprehension

Read this article for deeper understanding. No summary is required, although you may want to circle, underline, or mark key ideas and words.

About Experiments

In science, we need to make observations on various phenomena to form and test hypotheses. Some phenomena can be found and studied in nature, but scientists often need to create an experiment.

Experiments are tests under controlled conditions designed to demonstrate something scientists already know or to test something scientists wish to know.

Experiments vary greatly in their goal and scale, but **always rely on repeatable procedure and logical analysis of the results**. The process of designing and performing experiments is a part of the scientific method.

About the Scientific Method

The scientific method is the process used by scientists to acquire new knowledge and improve our understanding of the universe. It involves making observations on the phenomenon being studied, suggesting explanations for the observations, and testing these possible explanations, also called hypotheses, by making new observations. A hypothesis is a scientist's proposed explanation of a phenomenon which still must be tested.

Contrast of Scientific Theories and Laws

In science, a law is a mathematical relationship that exists between observations under a given set of conditions.

There is a fundamental difference between observations of the physical world and explanations of the nature of the physical world. Hypotheses and theories are explanations, whereas laws and measurements are observational.

Explanations	Observational
Theories & Hypothesis	Scientific Law

cK12.org

Draw Illustration

Copy and Label the Illustration in the Space Provided

cK12.org

Draw (Copy) the Illustration Here

Interpret a Graph

Write the title of the graph _____

Circle the type of chart this represents

    Bar Chart   Line Chart   Pie Chart   Other

If applicable,

    What does the X-axis represent _____

    What does the Y-axis imply _____

Summarize what this graph represents or conveys

_____

http://cen.acs.org

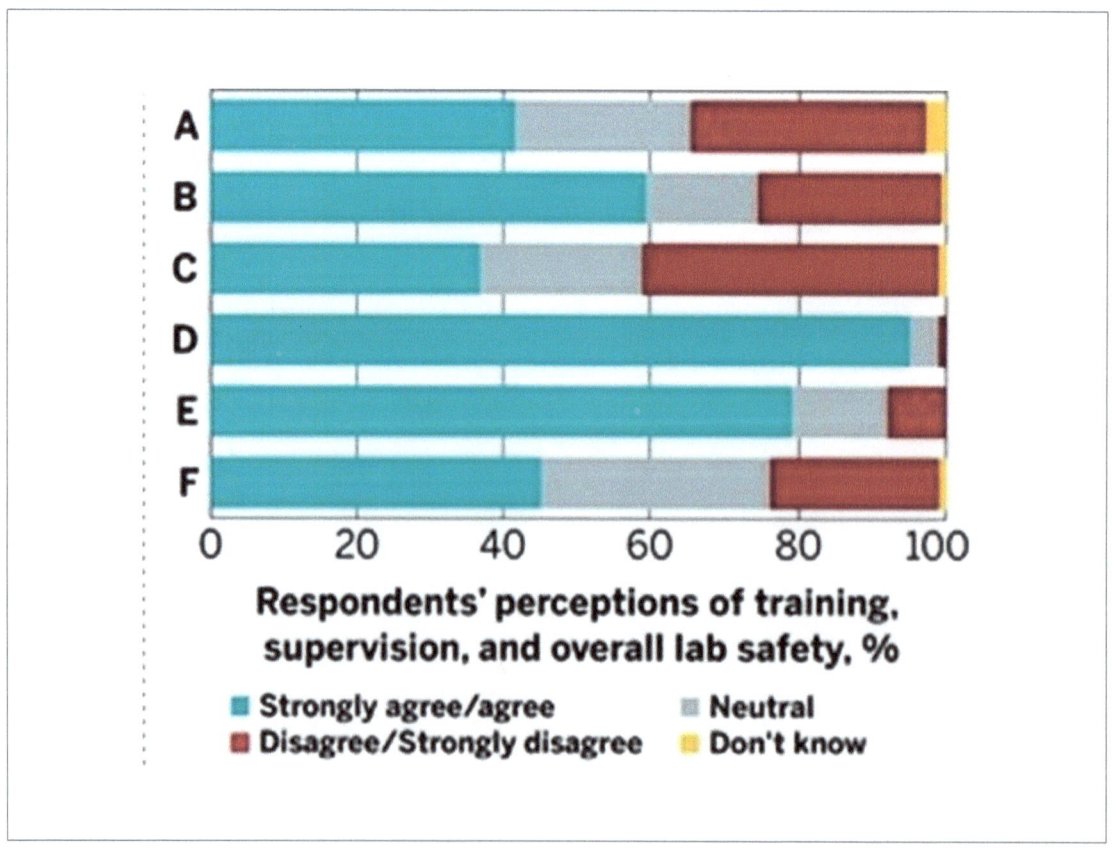

13.1 Connections Across Content

Summarize main points from each video.

Video Title / topic _____

Video Title / topic _____

Video Title / topic _____

Topic Introduction

Summarize your understanding of each paragraph.

Science is a logical activity. Science identifies, builds and organizes knowledge. Science knowledge can be tested. Scientists test ideas that other scientists have identified through observation. Through science, people can explain and predict things about the universe.

```
[                                                    ]
```

Modern science is broadly divided into the natural sciences and the social sciences. This class (earth science) along with physical science, chemistry, and biology are all natural sciences. Natural science studies and deals with the material world.

```
[                                                    ]
```

While this class is NOT a social science class, you may be interested to learn that social sciences deal with the study of people and societies. Psychology, sociology, anthropology, and history are among the many social sciences. (You do not need to recall this paragraph).

```
[                                                    ]
```

Natural science can be divided into two main branches: life science and physical science. You are studying earth science which will be presented primarily a "physical science" – the study of non-living things. But clearly, Earth has living things.

```
[                                                    ]
```

Read/Summarize Text

1. **Read the passage.**
2. **Underline key expressions in each sentence.**
3. **Re-write each word (or expression) you underlined.**
4. **Summarize the passage.**

Title of Passage.

> Earth science at our school is studied primarily as one of the physical sciences. Earth science is not studied deeply as a life science. Even so, there are some life science concepts that overlap with earth science. For example – erosion. When we study about erosion of the soil, some erosion is caused by non-living phenomena (like weather). But some erosion is caused by living things (plants and animals, for example).
>
> Earth science and chemistry have a lot of overlapping concepts. The Earth's material consists of matter made up of atoms and molecules. For example, minerals and rocks have a chemical composition.

Reference URL.

Re-write words you underlined

_____   _____   _____

_____   _____   _____

Using a complete sentence, summarize or rephrase the passage

Read Text for Comprehension

Read this article for deeper understanding. No summary is required, although you may want to circle, underline, or mark key ideas and words.

What is the difference between earth science and biology?

Earth sciences study different aspects of the planet such as weather, rocks, and soil. Biology studies life – such as animals and plants – living on Earth.

https://socratic.org

What is the difference between geology and earth science?

Geology is just one field within earth science that specifically studies rocks, their composition, and the processes that lead to the rocks and landforms on Earth. ... Earth science is a blanket term that includes geology as a subset; it also includes some aspects of biology, ecology, oceanography, meteorology, etc.

https://www.quora.com

Is Earth Science a Physical Science?

Physical science is the systematic study of the inorganic world, as distinct from the study of the organic world, which is the province of biological science. Physical science is ordinarily thought of as consisting of four broad areas: astronomy, physics, chemistry, and the Earth sciences.

https://www.britannica.com

Draw Illustration

Copy and Label the Illustration in the Space Provided

Illustration

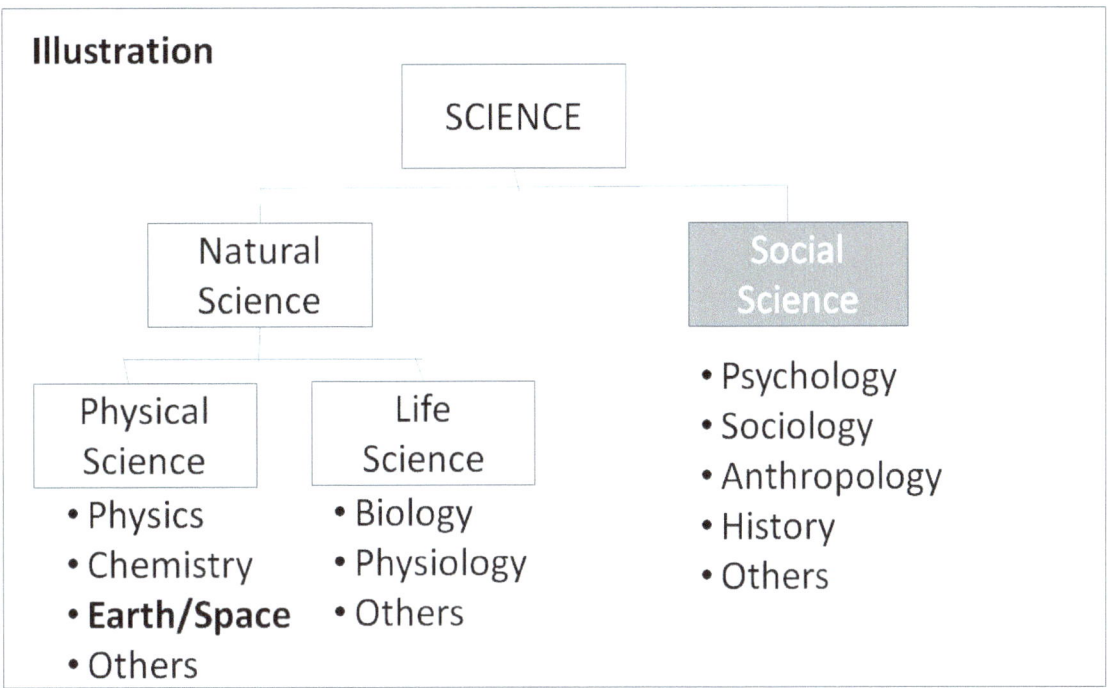

Reference URL.

Draw (Copy) the Illustration Here

Interpret a Graph

Write the title of the graph _____

Circle the type of chart this represents

    Bar Chart   Line Chart   Pie Chart   Other

If applicable,

    What does the X-axis represent _____

    What does the Y-axis imply _____

Summarize what this graph represents or conveys

_____

http://www.ielts-mentor.com

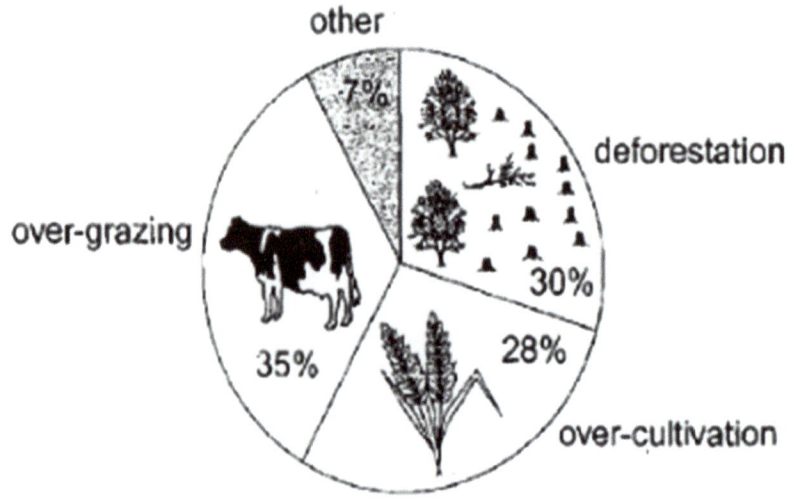

14.1 Maps

Summarize main points from each video.

Video Title / topic

Video Title / topic

Video Title / topic

Topic Introduction

Summarize your understanding of each paragraph.

Cartography is the study and practice of making maps. Combining science, aesthetics, and technique, cartography builds on the premise that reality can be modeled in ways that communicate spatial information effectively.

[ ]

Modern cartography constitutes many theoretical and practical foundations of geographic information systems.

[ ]

The earliest known map is a matter of some debate, both because the term "map" isn't well-defined and because some artifacts that might be maps might actually be something else.

[ ]

In ancient China, geographical literature dates to the 5th century BCE. The oldest extant Chinese maps come from the State of Qin, dated back to the 4th century BCE, during the Warring States period.

[ ]

Read/Summarize Text

1. **Read the passage.**
2. **Underline key expressions in each sentence.**
3. **Re-write each word (or expression) you underlined.**
4. **Summarize the passage.**

Fundamental problems of traditional cartography.

> **1.** Set the map's agenda and select traits of the object to be mapped. This is the concern of map editing. Traits may be physical, such as roads or land masses, or may be abstract, such as toponyms or political boundaries. **2.** Represent the terrain of the mapped object on flat media. This is the concern of map projections. **3.** Eliminate characteristics of the mapped object that are not relevant to the map's purpose. This is the concern of generalization. **4.** Reduce the complexity of the characteristics that will be mapped. This is also the concern of generalization. **5.** Orchestrate the elements of the map to best convey its message to its audience. This is the concern of map design.

Reference URL.

Re-write words you underlined

_____   _____   _____

_____   _____   _____

Using a complete sentence, summarize or rephrase the passage

Read Text for Comprehension

Read this article for deeper understanding. No summary is required, although you may want to circle, underline, or mark key ideas and words.

Cartography majors learn how to make maps. They study math, computer, and other techniques, including the interpretation of aerial photographs and remote-sensing data.

Cartographers and photogrammetrists collect, measure, and interpret geographic information in order to create and update maps and charts for regional planning, education, emergency response, and other purposes.

Work Environment

Although cartographers and photogrammetrists spend much of their time in offices, certain jobs require extensive travel to locations that are being mapped.

How to Become a Cartographer or Photogrammetrist

A bachelor's degree in cartography, geography, geomatics (the discipline that combines the science, engineering, math, and art of collecting and managing geographically referenced information), or surveying is the most common path of entry into this occupation. Cartographers and photogrammetrists must be licensed in some states.

Pay

The median annual wage for cartographers and photogrammetrists was $62,750 in May 2016.

Job Outlook

Employment of cartographers and photogrammetrists is projected to grow 29 percent from 2014 to 2024, much faster than the average for all occupations. The increasing use of maps for government planning should fuel employment growth. For this reason, job prospects are likely to be excellent.

Draw Illustration

Copy the Illustration in the Space Provided

http://map.asherrard.us/usa-map-outline/

Draw (Copy) the Illustration Here

Interpret a Graph

Write the title of the graph _____

Circle the type of chart this represents

    Bar Chart   Line Chart   Pie Chart   Other

If applicable,

    What does the X-axis represent _____

    What does the Y-axis imply _____

Summarize what this graph represents or conveys

_____

http://www.cccblog.org

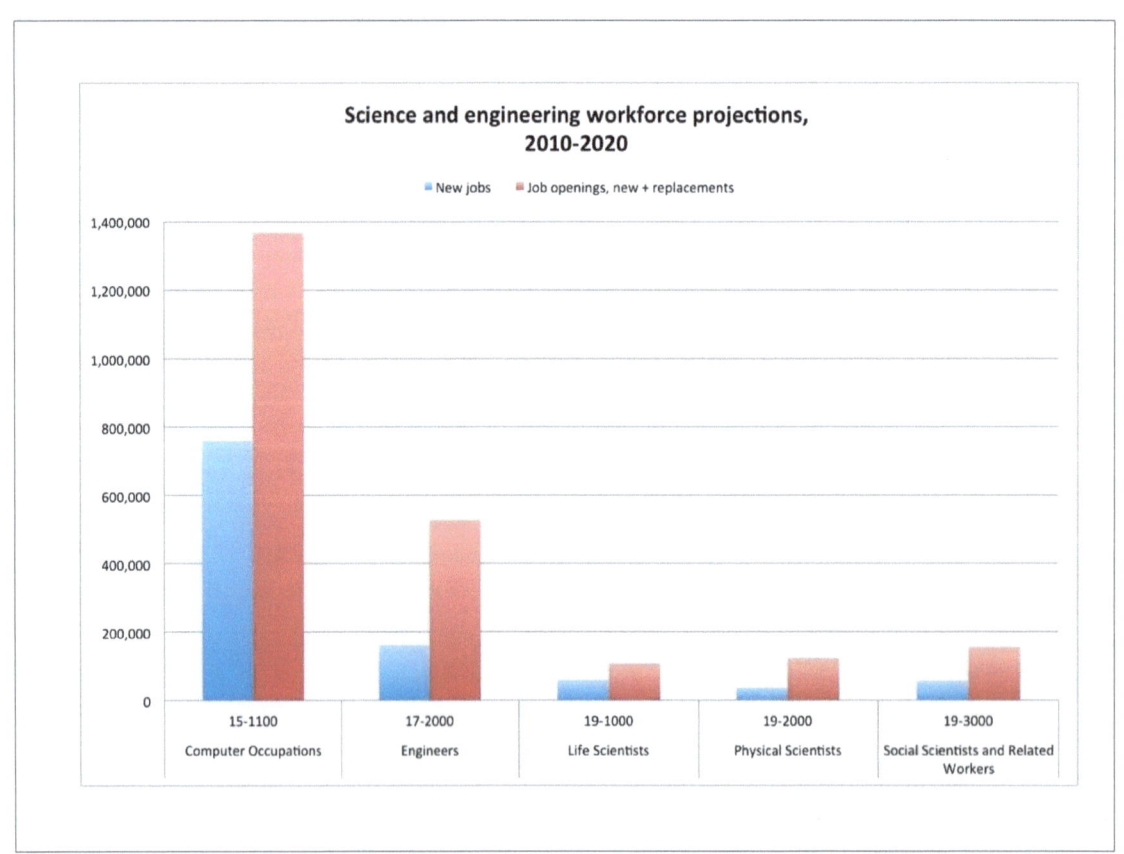

16.1 Earth's Chemistry

Summarize main points from each video.

Video Title / topic

Video Title / topic

Video Title / topic

Topic Introduction

Summarize your understanding of each paragraph.

There are ninety-two elements found on Earth. But only a few are very common. Most elements are found only in small quantities.

http://science.jrank.org/kids/pages/212/Common-Elements.html

[ ]

The Earth's atmosphere is primarily made up of nitrogen and oxygen. The Earth's oceans are made up primarily of water (hydrogen and oxygen) along with a dissolved ions (originating from sodium, chlorine, calcium, magnesium, and sulfur).

[ ]

The Earth's crust is mostly oxygen, silicon, aluminum, and iron. These, along with the additional elements of calcium, magnesium, potassium, and sodium make up well over 95% of the Earth's crust. Molecules and combinations of these elements make rocks and minerals.

[ ]

The Earth's core is primarily Iron, using the symbol of "Fe" on the Periodic Table. Iron, along with the metal Nickel make up the majority of the Earth's core.

[ ]

Read/Summarize Text

1. **Read the passage.**
2. **Underline key expressions in each sentence.**
3. **Re-write each word (or expression) you underlined.**
4. **Summarize the passage.**

Earth Chemistry at a Glance for Earth Day

> As scientists are not able to visit the Earth's deep interior or place instruments within it, they explore in subtle ways. One approach is to study the Earth with non-material probes, such as seismic waves emitted by earthquakes. As seismic waves pass through the Earth, they undergo sudden changes in direction and velocity at certain depths. These depths mark the major boundaries, also called discontinuities, that divide the Earth into crust, mantle and core.

www.decodedscience.org

Re-write words you underlined

_____   _____   _____

_____   _____   _____

Using a complete sentence, summarize or rephrase the passage

27

Read Text for Comprehension

Read this article for deeper understanding. No summary is required, although you may want to circle, underline, or mark key ideas and words.

The Crust. The Earth's crust is the thin outermost layer of the Earth, with an average depth of 24 km (15 mi). The crust accounts for 1.05% of the Earth's volume and 0.5% of its mass. The chemical elements oxygen, silicon and aluminum dominate the crustal composition. The major mineral type – the feldspars – are alumino-silicates of the alkali and alkaline-earth metals. Silicon dioxide is the second most common group.

The Mantle. The mantle extends from the base of the crust to the core and is about 2865 km (1780 mi) thick, occupying about 82.5% of the Earth's volume. The upper mantle is rich in olivine and pyroxenes. The major mineral type in the lower mantle appears to be pyroxenes, especially magnesium silicate. Scientists think that the lowest layer of the mantle called "D layer" is richer in aluminum and calcium than the higher layers of the mantle.

The Core. The core extends from the base of the mantle to the Earth's center, and is 6964 kn (4327 mi) in diameter – accounting for only 16.3% of the Earth's volume, but 33.5% of its mass. The core is made up of two distinct parts – a liquid outer core, which is 2260 km (1404 mi) thick, and a solid inner core, which has a radius of 1222 km (759 mi). The core is chemically distinct from the mantle and contains about 89% iron and 6% nickel. The remaining 5% is made of lighter elements, possibly sulfur – but we cannot rule out the presence of oxygen and silicon, in light of a 2013 study published in Nature, which calls them "prime candidates" for the lighter elements in the Earth's core.

Draw Illustration

Copy and Label the Illustration in the Space Provided

Illustration

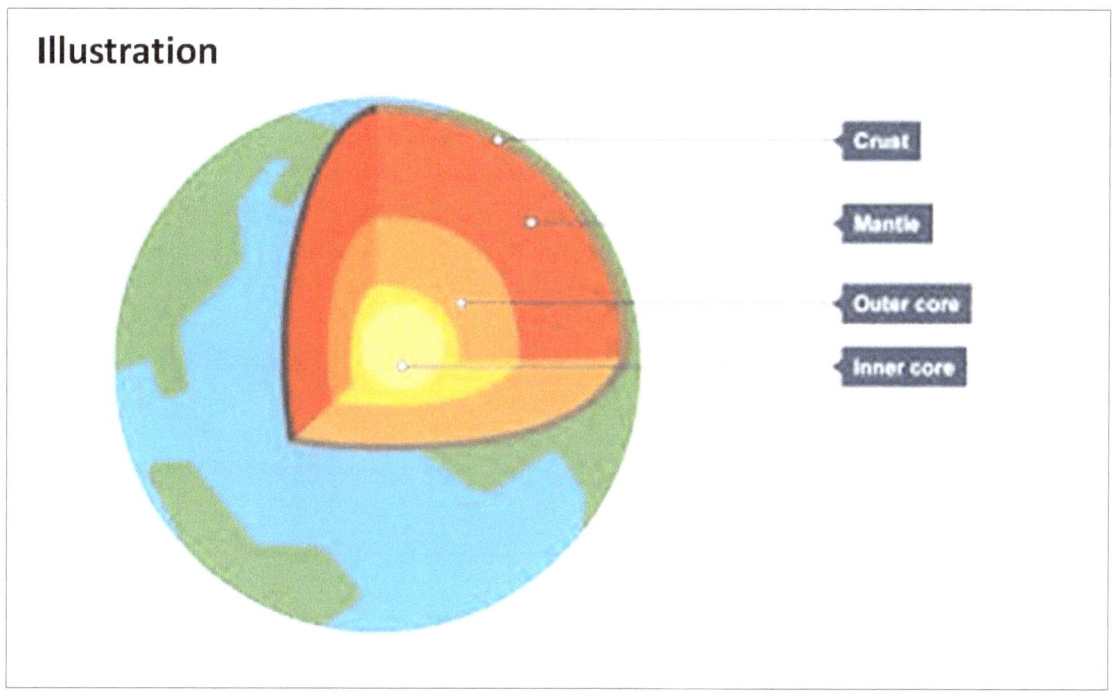

http://www.bbc.co.uk/education/guides/zysbgk7/revision

Draw (Copy) the Illustration Here

Interpret a Graph

Write the title of the graph _____

Circle the type of chart this represents
      Bar Chart   Line Chart   Pie Chart   Other

If applicable,
    What does the X-axis represent _____

    What does the Y-axis imply _____

Summarize what this graph represents or conveys

_____

http://science.jrank.org/kids/pages/212/Common-Elements.html

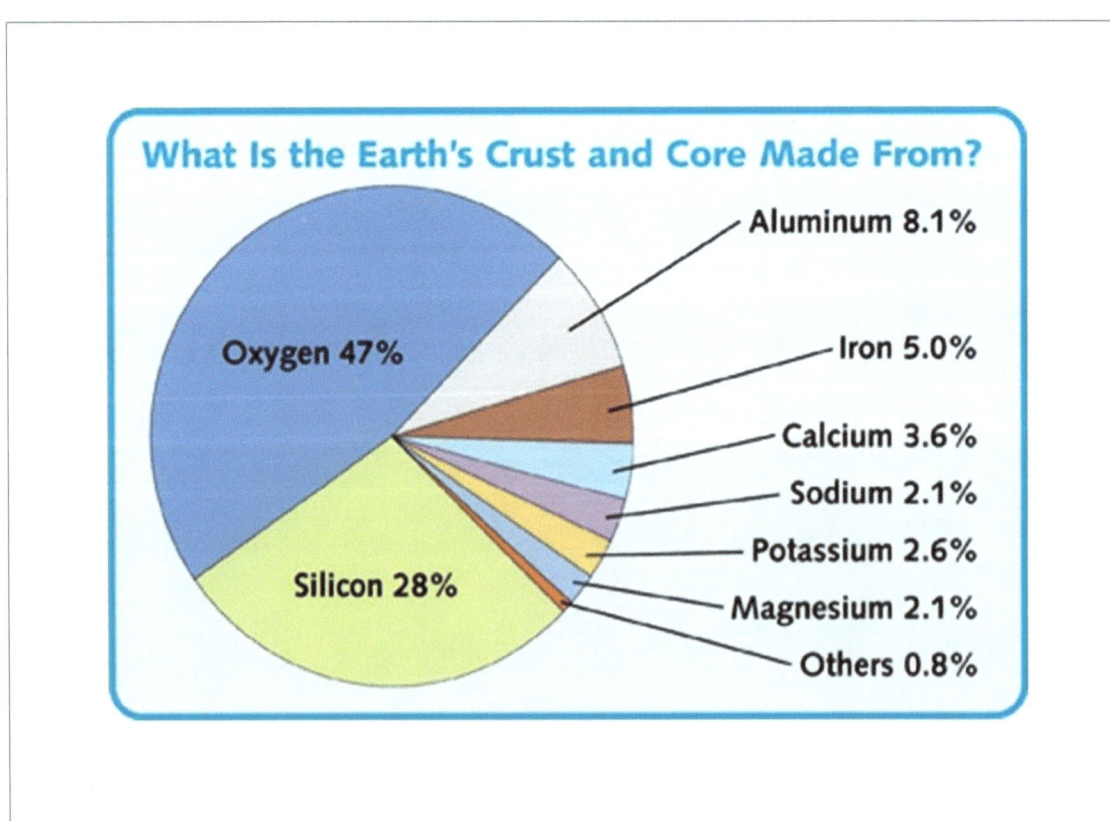

18.1 Types of rock

Summarize main points from each video.

Video Title / topic

Video Title / topic

Video Title / topic

Topic Introduction

Summarize your understanding of each paragraph.

There are three major types of rock, sedimentary, metamorphic, and igneous. They are all identified by their texture, streak, and location, among other factors..

[ ]

There is no agreed number of specific types of rocks. Any unique combination of chemical composition, mineralogy, grain size, texture, or other distinguishing characteristics can describe rock types.

[ ]

The rock cycle is a basic concept in geology that describes the time-consuming transitions through geologic time among the three main rock types: sedimentary, metamorphic, and igneous.

[ ]

Due to the driving forces of the rock cycle, plate tectonics and the water cycle, rocks do not remain in equilibrium and are forced to change as they encounter new environments.

[ ]

Wikipedia.com

Read/Summarize Text

1. **Read the passage.**
2. **Underline key expressions in each sentence.**
3. **Re-write each word (or expression) you underlined.**
4. **Summarize the passage.**

Rocks and Minerals.

> There are three types of rocks. They are all formed in different ways by nature. Just like minerals, rocks are solid and naturally forming. In fact, all rocks are made from two or more minerals. There are three different types of rocks, and all three form in different ways.
> - **Igneous** rocks are created when magma cools and hardens.
> - **Sedimentary** rocks form from the build-up of materials like the remains of plants or animals, minerals, and eroded fragments (pieces) of other rocks.
> - **Metamorphic** rocks start out as igneous or sedimentary rocks, but then they are transformed by extreme pressure or heat.

https://www.coolkidfacts.com/rocks-and-minerals/.

Re-write words you underlined

_____   _____   _____

_____   _____   _____

Using a complete sentence, summarize or rephrase the passage

33

Read Text for Comprehension

Read this article for deeper understanding. No summary is required, although you may want to circle, underline, or mark key ideas and words.

Igneous Rocks

Igneous rocks are formed from lava or magma. Magma is molten rock that is underground and lava is molten rock that erupts out on the surface. The two main types of igneous rocks are plutonic rocks and volcanic rocks. Plutonic rocks are formed when magma cools and solidifies underground. Volcanic rocks are formed from lava that flows on the surface of the Earth and other planets and then cools and solidifies.

The texture of an igneous rock depends on the size of the crystals in the rock. This tells us if the rock is plutonic or volcanic. When magma cools underground, it cools very slowly and when lava cools above ground, it cools quickly. When magma and lava cool, mineral crystals start to form in the molten rock. Plutonic rocks, which cool slowly underground, have large crystals because the crystals had enough time to grow to a large size. Volcanic rocks, which cool quickly above ground, have small crystals because the crystals did not have enough time to grow very large.

The type of igneous rock is also dependent on its composition (the elements that are present). There are many different compositions of magma and lava. Fortunately, most igneous rocks are one of three basic compositions:

- Felsic igneous rocks contain relatively high amounts of silicon, sodium, aluminum, potassium (Si, Na, Al, and K) and relatively low amounts of iron, magnesium, and calcium (Fe, Mg, Ca).

- Mafic rocks contain relatively low amounts of silicon, sodium, aluminum, potassium (Si, Na, Al, and K) and relatively high amounts of iron, magnesium, and calcium (Fe, Mg, Ca).

- Intermediate rocks are what their name sounds like. Their composition is in-between mafic and felsic rocks.

The different elements present in the different igneous compositions will form different minerals. Rocks with high amounts of iron (Fe) tend to form minerals that are dark in color (such as olivine and pyroxene). As result, mafic rocks tend to be dark in color and felsic rocks tend to be lighter in color. An example of a mafic rock is basalt, the black rock that forms from lava flows in places like Hawaii. An example of a felsic rock is granite, the light colored rocks that we find in places like the Sierra Nevada mountains in California.

http://ratw.asu.edu/aboutrocks_igneous.html

Draw Illustration

Copy and Label the Illustration in the Space Provided

Illustration

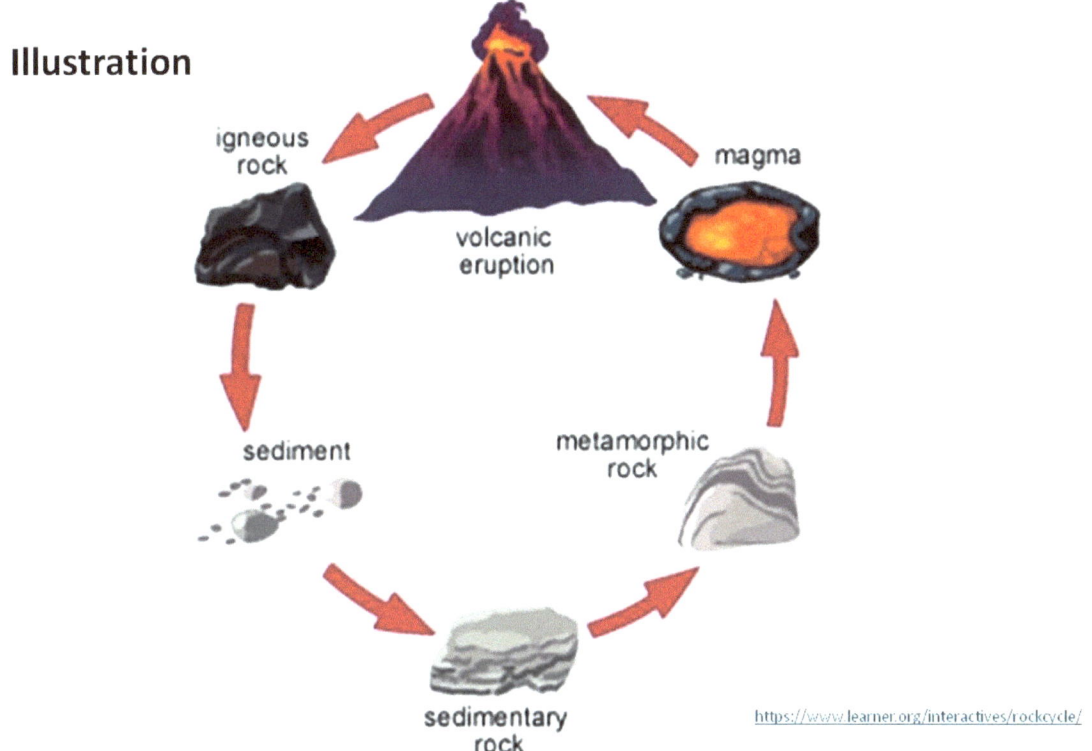

https://www.learner.org/interactives/rockcycle/

Draw (Copy) the Illustration Here

Interpret a Graph

Write the title of the graph _____

Circle the type of chart this represents

      Bar Chart   Line Chart   Pie Chart   Other

If applicable,
    What does the X-axis represent _____

    What does the Y-axis imply _____

Summarize what this graph represents or conveys

_____

Hester Williamson © 2017 SlidePlayer.com Inc.

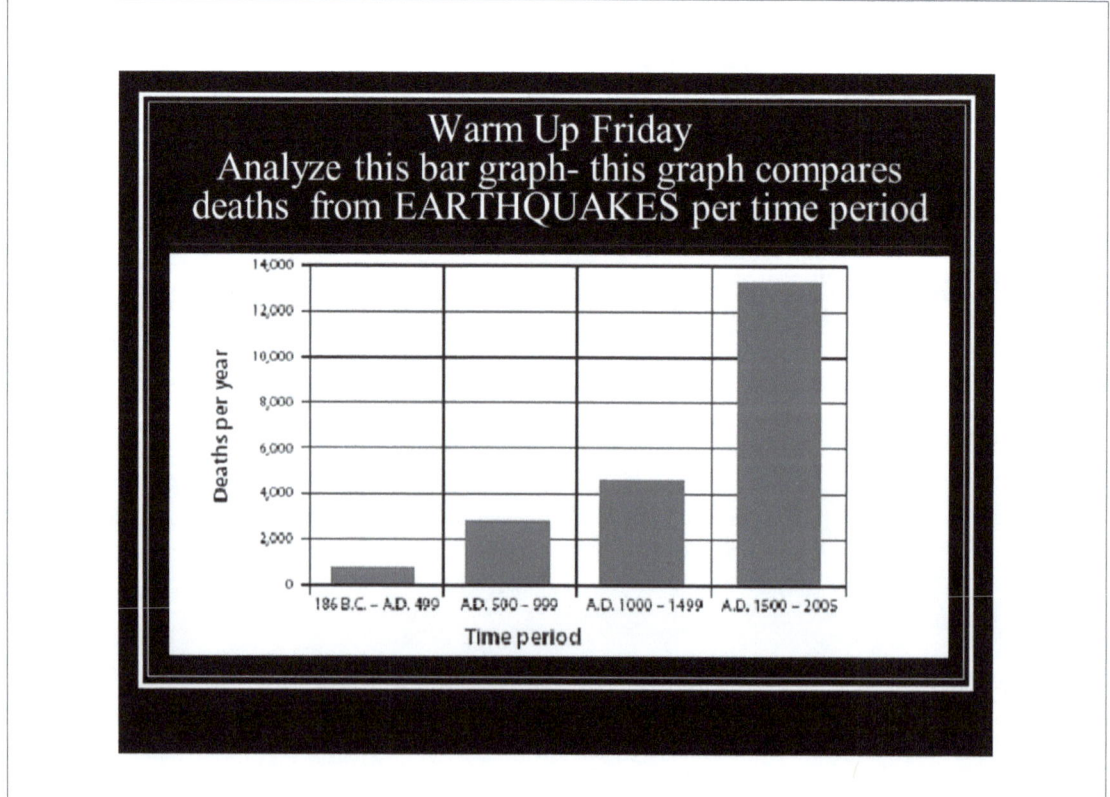

19.1 Resources and Energy

Summarize main points from each video.

Video Title / topic

Video Title / topic

Video Title / topic

Topic Introduction

Summarize your understanding of each paragraph.

Natural resources are resources that exist without actions of humankind. Some natural resources such as sunlight and air can be found everywhere, and are known as ubiquitous resources.

>

The vast majority of resources are theoretically exhaustible, which means they have a finite quantity and can be depleted if managed improperly.

>

Natural resources can be categorized as either renewable or non-renewable. Renewable resources can be replenished naturally. They replenish somewhat easily. Non-renewable resources form slowly or do not naturally form in the environment.

>

In recent years, the depletion of natural resources has become a major focus of governments and many other organizations. There is particular concern for rainforest regions which hold most of the Earth's biodiversity.

>

https://en.wikipedia.org/wiki/Natural_resource

Read/Summarize Text

1. **Read the passage.**
2. **Underline key expressions in each sentence.**
3. **Re-write each word (or expression) you underlined.**
4. **Summarize the passage.**

Natural Resources – and Management of Resources

Natural resource management is a discipline in the management of natural resources such as land, water, soil, plants and animals, with a particular focus on how management affects the quality of life for both present and future generations.

Hence sustainable development can be followed where there is a judicial use of resources which compromises the needs of the present generations as well as the future generations.

Management of natural resources involves identifying who has the right to use the resources and who does not.

https://en.wikipedia.org/wiki/Natural_resource

Re-write words you underlined

_____   _____   _____

_____   _____   _____

Using a complete sentence, summarize or rephrase the passage

Read Text for Comprehension

Read this article for deeper understanding. No summary is required, although you may want to circle, underline, or mark key ideas and words.

Renewable Energy Sources

Renewable Energy uses energy resources and technologies that are "clean" or "green" because they produce few if any pollutants. Many people use the terms "Alternative Energy", "Renewable Energy" and even "Green Energy" together in the same sentence when taking about energy sources as though they all mean the same thing, but they are not the same. Each term means something different when talking about energy systems. So what does renewable energy mean.

Some say that alternative energy comprises everything that is not based on fossil fuel consumption. While these may be alternative energy sources compared to conventional fossil fuels, alternative energy in its broadest sense, is any type of energy that replaces another, so we can correctly say that coal energy is an alternative energy source compared to crude oil or natural gas but as we now know, coal is a fossil fuel and burning it is bad for the environment. Even nuclear energy was once considered to be an "alternative" to conventional fossil fuels, and was thus called an alternative energy source.

Renewable Energy on the other hand uses renewable energy sources that are continually replenished by Mother nature producing a usable energy that can not be used up faster that it is consumed. These energy sources created mainly by the Sun shinning on the Earth are converted into different forms, such as: solar radiation to wind or water based energy which is distributed over the Earth and atmosphere, the Earth's geothermal heat, and plants in the form of biomass. Renewable energy technologies turn these fuels into usable forms of energy, most often electricity, but also heat, chemicals, or mechanical power. So what are renewable resources.

http://www.alternative-energy-tutorials.com/energy-articles/renewable-energy-sources-a-brief-summary.html

Draw Illustration

Copy and Label the Illustration in the Space Provided

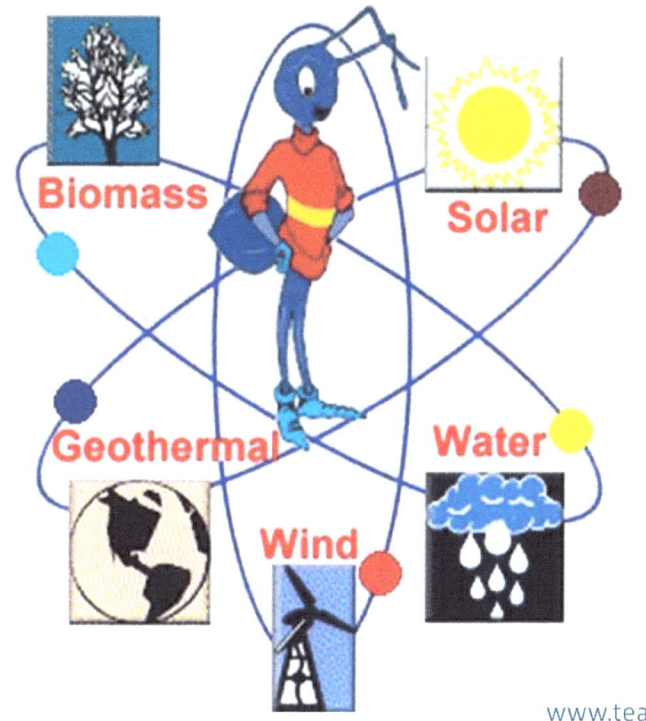

www.teachengineering.org

Draw (Copy) the Illustration Here

Interpret a Graph

Write the title of the graph _____

Circle the type of chart this represents
      Bar Chart   Line Chart   Pie Chart   Other

If applicable,
    What does the X-axis represent _____

    What does the Y-axis imply _____

Summarize what this graph represents or conveys

_____

https://www.eia.gov/renewable/

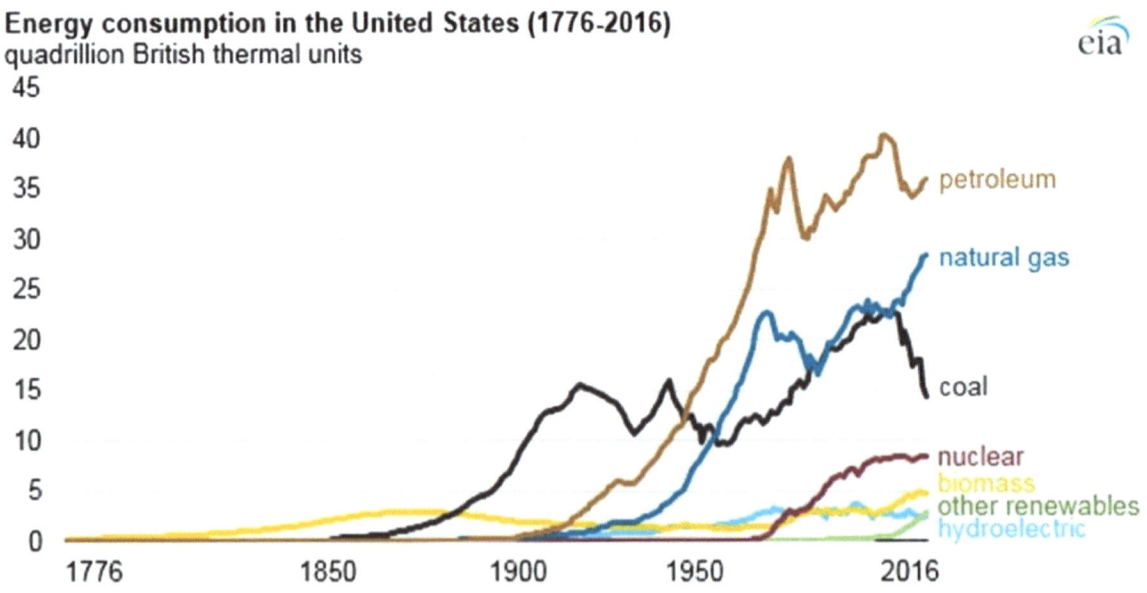

21.1 Early Earth – The Rock Record

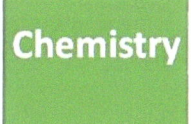

Summarize main points from each video.

Video Title / topic

Video Title / topic

Video Title / topic

Topic Introduction

Summarize your understanding of each paragraph.

Earth is about 4.4 billion years old. Among other forms of evidence, scientists confirmed the age of an individual atom of lead contained in a tiny zircon crystal. To-date, the crystal is the oldest rock fragment ever found on Earth — 4.375 billion years old.

Geologists have carefully sorted out more than 100,000 microscopic Jack Hills zircons that date back to Earth's early epochs, from 3 billion to nearly 4.4 billion years ago. The crystals contain microscopic inclusions, providing a window into conditions on early Earth.

The geologic record in stratigraphy, paleontology and other natural sciences refers to the entirety of the layers of rock strata — deposits laid down by volcanism or by deposition of sediment derived from weathering detritus (clays, sands etc.) including all its fossil content.

At a certain locality on the Earth's surface, the rock column provides a cross section of the natural history in the area during the time covered by the age of the rocks. This is sometimes called the rock history and gives a window into the natural history of the location.

https://en.wikipedia.org/wiki/Geologic_record
https://www.livescience.com/

Read/Summarize Text

1. **Read the passage.**
2. **Underline key expressions in each sentence.**
3. **Re-write each word (or expression) you underlined.**
4. **Summarize the passage.**

Wikipedia - Stratigraphy.

Stratigraphy is a branch of geology concerned with the study of rock layers (strata) and layering (stratification). It is primarily used in the study of sedimentary and layered volcanic rocks. Stratigraphy has two related subfields: lithologic stratigraphy or lithostratigraphy, and biologic stratigraphy or biostratigraphy.

Variation in rock units, most obviously displayed as visible layering, is due to physical contrasts in rock type (lithology). This variation can occur vertically as layering (bedding), or laterally, and reflects changes in environments of deposition (known as facies change).

https://en.wikipedia.org/wiki/Stratigraphy

Re-write words you underlined

_____   _____   _____

_____   _____   _____

Using a complete sentence, summarize or rephrase the passage

Read Text for Comprehension

Read this article for deeper understanding. No summary is required, although you may want to circle, underline, or mark key ideas and words.

Wikipedia: Stratum

In geology and related fields, a stratum (plural: strata) is a layer of sedimentary rock or soil, or igneous rock where formed at the earth's surface[1], with internally consistent characteristics that distinguish it from other layers. The "stratum" is the fundamental unit in a stratigraphic column and forms the basis of the study of stratigraphy.

Characteristics

Each layer is generally one of a number of parallel layers that lie one upon another, laid down by natural processes. They may extend over hundreds of thousands of square kilometers of the Earth's surface. Strata are typically seen as bands of different colored or differently structured material exposed in cliffs, road cuts, quarries, and river banks. Individual bands may vary in thickness from a few millimeters to a kilometer or more. Each band represents a specific mode of deposition: river silt, beach sand, coal swamp, sand dune, lava bed, etc.

Naming

Geologists study rock strata and categorize them by the material of beds. Each distinct layer is typically assigned to the name of sheet, usually based on a town, river, mountain, or region where the formation is exposed and available for study. For example, the Burgess Shale is a thick exposure of dark, occasionally fossiliferous, shale exposed high in the Canadian Rockies near Burgess Pass. Slight distinctions in material in a formation may be described as "members" (or sometimes "beds"). Formations are collected into "groups" while groups may be collected into "supergroups".

https://en.wikipedia.org/wiki/Stratum

Draw Illustration

Copy and Label the Illustration in the Space Provided

Grand Canyon's Three Sets of Rocks

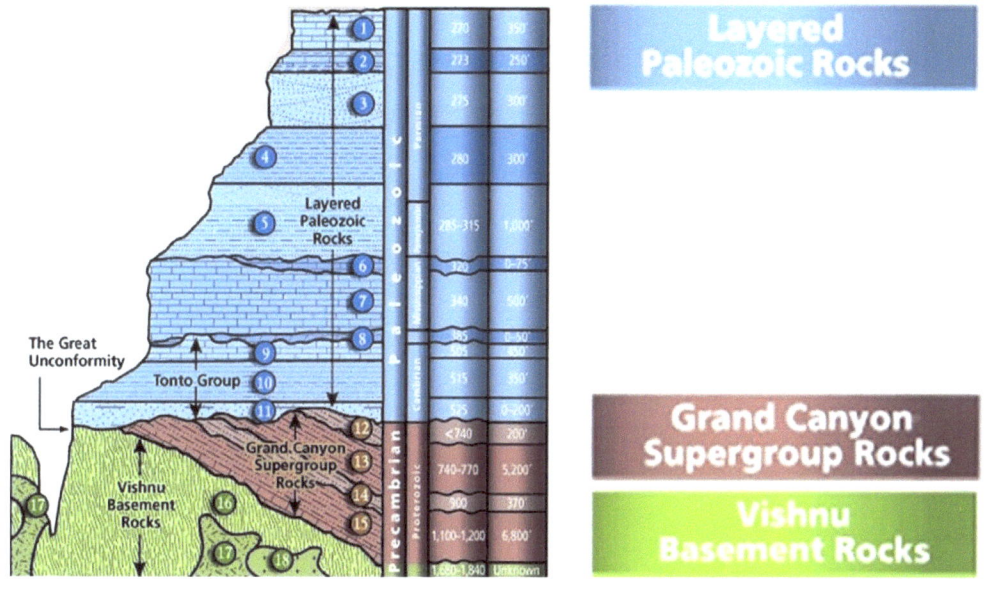

https://en.wikipedia.org/wiki/Great_Unconformity

Draw (Copy) the Illustration Here

Interpret a Chart

The Name of this Chart is "Geological Time Scale"

Discuss this chart with a teacher-assigned "shoulder-buddy" ... Summarize what this graph represents or conveys.

https://www.britannica.com/science/Holocene-Epoch

Geologic time scale

Eonothem/Eon	Erathem/Era	System/Period	Series/Epoch	Stage/Age	mya[1]
Phanerozoic	Cenozoic	Quaternary	Holocene		0.0117
			Pleistocene	Upper	0.126
				Middle	0.781
				Calabrian	1.80
				Gelasian	2.58
		Neogene	Pliocene	Piacenzian	3.600
				Zanclean	5.333
			Miocene	Messinian	7.246
				Tortonian	11.63
				Serravallian	13.82
				Langhian	15.97
				Burdigalian	20.44
				Aquitanian	23.03
		Paleogene	Oligocene	Chattian	28.1
				Rupelian	33.9
			Eocene	Priabonian	37.8
				Bartonian	41.2
				Lutetian	47.8
				Ypresian	56.0
			Paleocene	Thanetian	59.2
				Selandian	61.6
				Danian	66.0
	Mesozoic	Cretaceous	Upper	Maastrichtian	72.1 ± 0.2
				Campanian	83.6 ± 0.2
				Santonian	86.3 ± 0.5
				Coniacian	89.8 ± 0.3
				Turonian	93.9
				Cenomanian	100.5
			Lower	Albian	~113
				Aptian	~125.0
				Barremian	~129.4
				Hauterivian	~132.9
				Valanginian	~139.8
				Berriasian	~145.0

Eonothem/Eon	Erathem/Era	System/Period	Series/Epoch	Stage/Age	mya[1]
Phanerozoic	Mesozoic	Jurassic	Upper	Tithonian	~145.0
				Kimmeridgian	152.1 ± 0.9
				Oxfordian	157.3 ± 1.0
			Middle	Callovian	163.5 ± 1.0
				Bathonian	166.1 ± 1.2
				Bajocian	168.3 ± 1.3
				Aalenian	170.3 ± 1.4
			Lower	Toarcian	174.1 ± 1.0
				Pliensbachian	182.7 ± 0.7
				Sinemurian	190.8 ± 1.0
				Hettangian	199.3 ± 0.3
		Triassic	Upper	Rhaetian	201.3 ± 0.2
				Norian	~208.5
				Carnian	~228.0
			Middle	Ladinian	~235.0
				Anisian	~242.0
			Lower	Olenekian	247.2
				Induan	251.2
	Paleozoic	Permian	Lopingian	Changhsingian	252.2 ± 0.5
				Wuchiapingian	254.2 ± 0.1
			Guadalupian	Capitanian	259.9 ± 0.4
				Wordian	265.1 ± 0.4
				Roadian	268.8 ± 0.5
			Cisuralian	Kungurian	272.3 ± 0.5
				Artinskian	279.3 ± 0.6
				Sakmarian	290.1 ± 0.1
				Asselian	295.5 ± 0.4
		Carboniferous	Pennsylvanian[2] U	Gzhelian	298.9 ± 0.2
				Kasimovian	303.7 ± 0.1
			M	Moscovian	307.0 ± 0.1
			L	Bashkirian	315.2 ± 0.2
			Mississippian[2] U	Serpukhovian	323.2 ± 0.4
			M	Visean	330.9 ± 0.2
			L	Tournaisian	346.7 ± 0.4
					358.9 ± 0.4

Eonothem/Eon	Erathem/Era	System/Period	Series/Epoch	Stage/Age	mya[1]
Phanerozoic	Paleozoic	Devonian	Upper	Famennian	358.9 ± 0.4
				Frasnian	372.2 ± 1.6
			Middle	Givetian	382.7 ± 1.6
				Eifelian	387.7 ± 0.8
			Lower	Emsian	393.3 ± 1.2
				Pragian	407.6 ± 2.6
				Lochkovian	410.8 ± 2.8
		Silurian	Pridoli		419.2 ± 3.2
			Ludlow	Ludfordian	423.0 ± 2.3
				Gorstian	425.6 ± 0.9
			Wenlock	Homerian	427.4 ± 0.5
				Sheinwoodian	430.5 ± 0.7
			Llandovery	Telychian	433.4 ± 0.8
				Aeronian	438.5 ± 1.1
				Rhuddanian	440.8 ± 1.2
		Ordovician	Upper	Hirnantian	443.8 ± 1.5
				Katian	445.2 ± 1.4
				Sandbian	453.0 ± 0.7
			Middle	Darriwilian	458.4 ± 0.9
				Dapingian	467.3 ± 1.1
			Lower	Floian	470.0 ± 1.4
				Tremadocian	477.7 ± 1.4
		Cambrian[3]	Furongian	Stage 10	485.4 ± 1.9
				Jiangshanian	~489.5
				Paibian	~494.0
			Series 3	Guzhangian	~497.0
				Drumian	~500.5
				Stage 5	~504.5
			Series 2	Stage 4	~509.0
				Stage 3	~514.0
			Terreneuvian	Stage 2	~521.0
				Fortunian	~529.0
					541.0 ± 1.0

Eonothem/Eon	Erathem/Era	System/Period	mya[1]
Precambrian	Proterozoic	Neoproterozoic	Ediacaran	~541.0 ± 1.0
			Cryogenian	~635
			Tonian	~720
		Mesoproterozoic	Stenian	1,000
			Ectasian	1,200
			Calymmian	1,400
		Paleoproterozoic	Statherian	1,600
			Orosirian	1,800
			Rhyacian	2,050
			Siderian	2,300
	Archean	Neoarchean		2,500
		Mesoarchean		2,800
		Paleoarchean		3,200
		Eoarchean		3,600
	Hadean[4]			4,000
				~4,600

[1] Millions of years ago.
[2] Both the Mississippian and Pennsylvanian time units are formally designated as subperiods within the Carboniferous Period.
[3] Several Cambrian unit age boundaries are informal and are awaiting ratified definitions.
[4] The Hadean Eon is an informal interval of geologic time.

Published with permission from the International Commission on Stratigraphy (ICS). International chronostratigraphic units, ranks, names, and formal status are approved by the ICS and ratified by the International Union of Geological Sciences (IUGS).
Source: 2015 International Chronostratigraphic Chart produced by the ICS.

23.1 Plate Tectonics – Today's Earth

Summarize main points from each video.

Video Title / topic _____

Video Title / topic _____

Video Title / topic _____

Topic Introduction

Summarize your understanding of each paragraph.

Plate tectonics is a scientific theory describing the large-scale motion of seven large plates and the movements of a larger number of smaller plates of the Earth's lithosphere.

[ ]

The lithosphere is the rigid outermost shell of a planet (the crust and upper mantle). The lithosphere is broken into tectonic plates. The Earth's lithosphere is composed of seven major plates and many minor (smaller) plates.

[ ]

Tectonic plates are composed of oceanic lithosphere and thicker continental lithosphere, each topped by its own kind of crust.

[ ]

Where the plates meet, their relative motion determines the type of boundary: convergent, divergent, or transform. Earthquakes, volcanic activity, mountain-building, and oceanic trench formation occur along these plate boundaries (or faults).

[ ]

Adapted from: https://en.wikipedia.org/wiki/Plate_tectonics

Read/Summarize Text

1. **Read the passage.**
2. **Underline key expressions in each sentence.**
3. **Re-write each word (or expression) you underlined.**
4. **Summarize the passage.**

Historical development of the Theory of Plate Tectonics

In line with other previous and contemporaneous proposals, in 1912 the meteorologist Alfred Wegener amply described what he called continental drift, expanded in his 1915 book The Origin of Continents and Oceans and the scientific debate started that would end up fifty years later in the theory of plate tectonics.

History of Ocean Basins was published in 1962 and explained the mechanism behind Alfred Wegener's continental drift theory. In the paper Hess described how hot magma would rise from under the crust at the Great Global Rift. When the magma cooled, it would expand and push the tectonic plates apart.

Adaptations for wikipedia and psu.edu.

Re-write words you underlined

_____   _____   _____

_____   _____   _____

Using a complete sentence, summarize or rephrase the passage

Read Text for Comprehension

Read this article for deeper understanding. No summary is required, although you may want to circle, underline, or mark key ideas and words.

DEFINITION - Plate tectonics is the theory explaining the movement of the earth's plates and the processes that occur at their boundaries.

PLATES - Plates are variously-sized (approximately 60 miles thick) areas of the earth's crust and mantle (also called the lithosphere) that move slowly around the mantle's asthenosphere and are predominantly responsible for the earth's volcanoes and earthquakes. The asthenosphere is a portion of the mantle that consists of extremely hot, plastic-like rock that is partially melted.

DIVERGENT PLATE BOUNDARY - The lithospheric, the Earth's crust and upper mantle, includes three plate boundaries, the first of which is a divergent plate boundary. On a divergent plate boundary, the plates move apart in opposite directions.

CONVERGENT PLATE BOUNDARY - A mountain can be formed by convergent boundaries. On the second type of boundary, a convergent boundary, the plates are pushed together. Convergent plate boundaries help create mountains and volcanoes.

TRANSFORM FAULT - The third type of plate boundary is a transform fault. On a transform fault, the plates move in opposite but parallel directions along a fracture. In other words, the plates slide past one another.

THE EARTH'S CORE - The innermost part of the Earth is called the core. The core is extremely hot (4,300 degrees Celsius) and is made mostly of iron. The core is mostly solid but is surrounded by a liquid molten material.

THE EARTH'S MANTLE - The thickest of the Earth's three zones, the mantle surrounds the core and is mostly solid rock. A small portion of the mantle, the asthenosphere, is very hot (approximately 3,700 degrees Celsius), partially melted rock.

THE EARTH'S CRUST - The oceanic crust extends below the ocean floor. The Earth's crust is the outermost and thinnest layer of the Earth's three zones. It consists of the continental and oceanic crusts.

CONVECTION CELLS - Convection cells are believed to be what helps keep the plates moving. The plates rest on the constantly moving, plastic-like rock of the lower mantle (asthenosphere) and move in a similar fashion to convection in the atmosphere.

CONTINENTAL DRIFT - The theory of plate tectonics developed in the 1960s from an earlier theory called continental drift. Continental drift was introduced by Alfred Lothar Wegener in 1912, and it claims that the continents were once connected and that they gradually drifted apart over millions of years. Plate tectonics is significant because it explains how continental drift can occur.

https://sciencing.com/10-plate-tectonics-7714571.html

Draw Illustration

Copy and Label the Illustration in the Space Provided

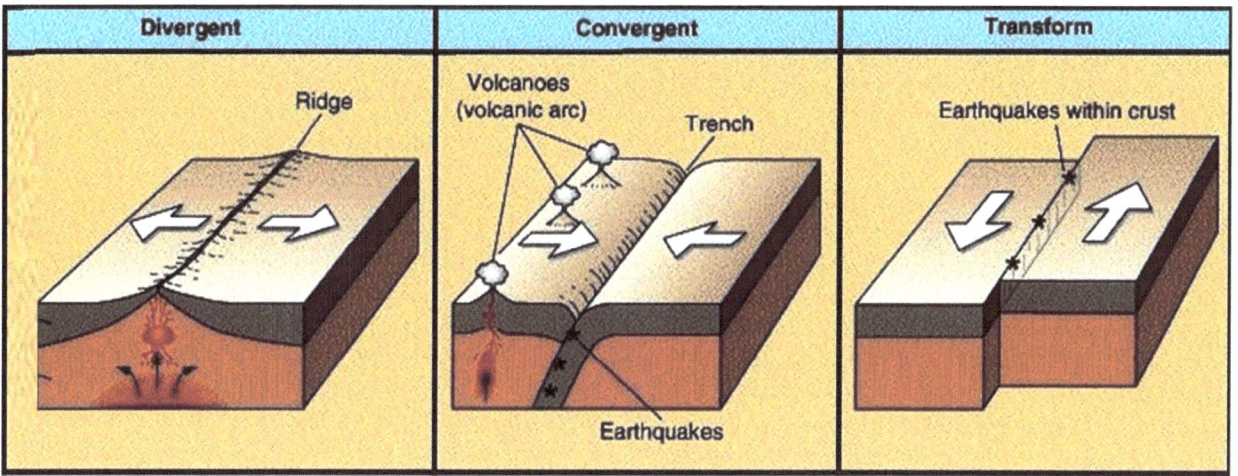

Adapted from: https://www.geologyin.com/2016/05/12-facts-you-should-know-about-plate.html

Draw (Copy) the Illustration Here

Interpret a Graphic

Write the title of the graphic
_____

Summarize what this graph represents or conveys
_____

In what way does this "Life Science" graphic correspond to the information presented in this Student Handout?

http://mayakg.blogspot.com/2016/03/geologic-timeline-reflection.html

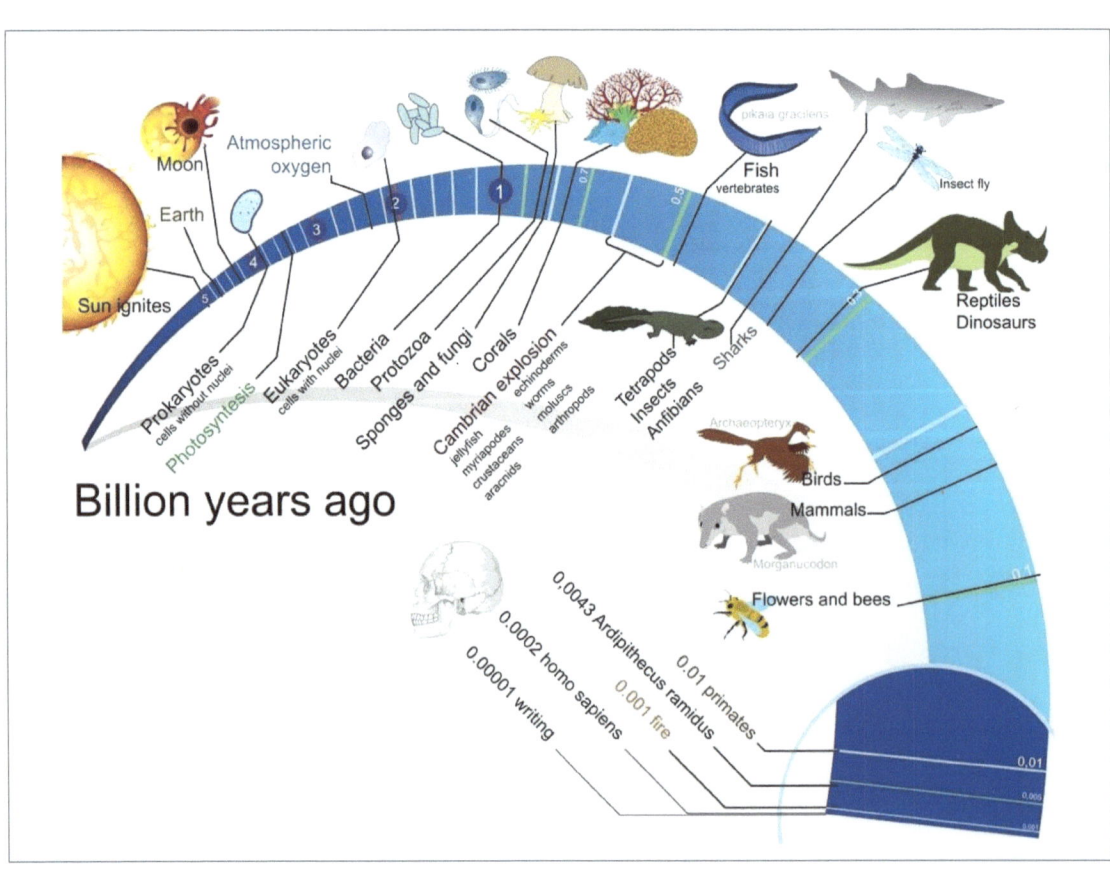

24.1 Deformation of the Crust

Summarize main points from each video.

Video Title / topic

Video Title / topic

Video Title / topic

Topic Introduction

Summarize your understanding of each paragraph.

Earth continually changes. Earth's crust undergoes change. The expressions "deformation" and "deform" are commonly used in describing much of the Earth's change and reshaping. Deformation is the bending, tilting, and breaking of Earth's crust.

Deformation is the change in shape of rock in response to stress. There are a dozen or so words like "deformation" and "stress" that are commonly used in every-day language. In science, though, these words often have a specific meaning.

The crust is broken into several parts, known as the continental plates. When the plates are pulled or pushed together, stress occurs. Four types of stresses affect the Earth's crust: compression, tension, shear and confining (sometimes called "overburden") stress.

Take note of differences between the specific (scientific) meaning of words in this topic – compared to other meanings in every day language. For examples, "stress" has several meanings. Some other words to pay attention to are "strain", "fault", "strike", and "fold."

Read/Summarize Text

1. **Read the passage.**
2. **Underline key expressions in each sentence.**
3. **Re-write each word (or expression) you underlined.**
4. **Summarize the passage.**

Stress, strain, and structures in rock.

> Tension, compression, and shearing work over millions of years to change the shape and volume of rock. Stress that pushes masses of rock in opposite directions, in a sideways movement. Strain can cause rock to break and slip apart or to change its shape. The differences between stress, strain and structures formed during strain become key concepts.
>
> - **Stress** is a force acting on a rock per unit area.
>
> - **Structures** in geology are deformation features that result from permanent (brittle or ductile) **strain.**

https://serc.carleton.edu

Re-write words you underlined

_____   _____   _____

_____   _____   _____

Using a complete sentence, summarize or rephrase the passage

Read Text for Comprehension

Read this article for deeper understanding. No summary is required, although you may want to circle, underline, or mark key ideas and words.

Rocks deform

Many students have a difficult time realizing that rocks can bend or break. They also may have difficulty imagining the forces necessary to fold or fault rocks or comprehending that the seemingly constant Earth can change dramatically over time.

This is especially true of students who live in tectonically stable areas. If students are to understand the basics of stress and strain, they must overcome this barrier since it will be difficult to examine the causes and conditions of deformation if students cannot comprehend deformation.

Stress causes strain, strain results in structures

Many geologists consider it important for introductory students to understand that visible structures are a record of the stress and physical conditions in the Earth. As a result, the differences between stress, strain and structures formed during strain become key concepts.

Stress, strain and structure start with the same three letters, yet mean very different things. These words are also used differently in geology than in common usage in English, which can cause confusion. However, here are some tricks that I use to remember:

- Stress is the same as pressure. When you are under pressure, you are stressed!
- Stress can happen with out strain, but strain cannot happen without stress.

Different conditions lead to different deformation styles

There are many factors that contribute to the style of the deformation in a rock, including pressure, temperature, rock composition, presence or absence of fluids, type of stress, rate of stress, and others. However, the type of stress, the rate of stress and the temperature may be the most critical factors for most introductory students.

Relating faults to stress - hanging walls, footwalls, and different types of faults

One of the goals of structural geology is to relate the nature of deformation to the stress that caused it. Therefore, it is important that students be able to distinguish between normal faults (generated by tension) and reverse faults (generated by compression).

https://serc.carleton.edu/quantskills/methods/quantlit/stressandstrain.html#faults

Draw Illustration

Copy and Label the Illustration in the Space Provided

Force applied to an Area – Differences in the Effect

http://geologylearn.blogspot.com/2016/03/rock-deformation.html

Draw (Copy) the Illustration Here

Draw Illustration

Copy and Label the Illustration in the Space Provided

Differences in Brittle compared to Ductile deformation

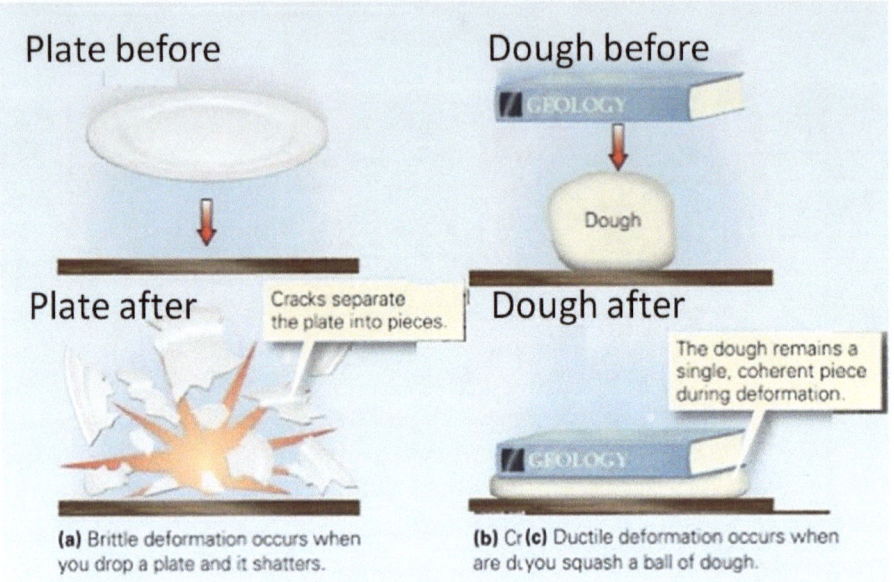

http://geologylearn.blogspot.com/2016/03/rock-deformation.html

Draw (Copy) the Illustration Here

25.1 Earthquakes Volcanoes & Tsunamis

Summarize main points from each video.

Video Title / topic

Video Title / topic

Video Title / topic

Topic Introduction

Summarize your understanding of each paragraph.

An earthquake is the shaking of the surface of the Earth, resulting from the sudden release of energy in the Earth's lithosphere. They range in size. The seismic activity of an area refers to the frequency, type and size of earthquakes experienced over a period of time.

[ ]

A volcano is a rupture in the crust of a planetary-mass object, such as Earth, that allows hot lava, volcanic ash, and gases to escape from a magma chamber below the surface. Earth's volcanoes occur because its crust is broken into 17 major, rigid tectonic plates.

[ ]

A tsunami (also known as a seismic sea wave) is a series of waves in a water body caused by displacement of a large volume of water, generally in an ocean or a large lake. Earthquakes, volcanic eruptions and underwater explosions have the potential to generate one.

[ ]

Each of these large-scale occurrences have an impact on the Earth's surface and life occupying that space. Combined, these have had a significant influence on the Earth – its landscape, its capacity to support life, and other crucial aspects - for billions of years.

[ ]

Wikipedia.com

Read/Summarize Text

1. **Read the passage.**
2. **Underline key expressions in each sentence.**
3. **Re-write each word (or expression) you underlined.**
4. **Summarize the passage.**

1994 Northridge earthquake

The 1994 Northridge earthquake occurred on January 17, had its epicenter in Reseda, a neighborhood in the north-central San Fernando Valley region of Los Angeles, California. It had a duration of approximately 10–20 seconds. The blind thrust earthquake had a moment magnitude of 6.7, which produced ground acceleration that was the highest ever instrumentally recorded in an urban area in North America.

It was felt as far away as Las Vegas, Nevada, about 220 miles from the epicenter. The death toll was 57, with more than 8,700 injured. In addition, property damage was estimated to be between $13 and $50 billion, making it one of the costliest natural disasters in U.S. history.

https://en.wikipedia.org/wiki/1994_Northridge_earthquake

Re-write words you underlined

_____   _____   _____

_____   _____   _____

Using a complete sentence, summarize or rephrase the passage

Read Text for Comprehension

Read this article for deeper understanding. No summary is required, although you may want to circle, underline, or mark key ideas and words.

2011 Tōhoku earthquake and tsunami

The 2011 earthquake off the Pacific coast of Tōhoku was a magnitude 9.0–9.1 (Mw) undersea megathrust earthquake off the coast of Japan that occurred at 14:46 JST (05:46 UTC) on Friday 11 March 2011, with the epicentre approximately 70 kilometres (43 mi) east of the Oshika Peninsula of Tōhoku and the hypocenter at an underwater depth of approximately 29 km (18 mi). The earthquake is often referred to in Japan as the Great East Japan Earthquake and is also known as the 2011 Tōhoku earthquake, and the 3.11 earthquake. It was the most powerful earthquake ever recorded in Japan, and the fourth most powerful earthquake in the world since modern record-keeping began in 1900.

The earthquake triggered powerful tsunami waves that reached heights of up to 40.5 metres (133 ft) in Miyako in Tōhoku's Iwate Prefecture, and which, in the Sendai area, traveled up to 10 km (6 mi) inland. The earthquake moved Honshu (the main island of Japan) 2.4 m (8 ft) east, shifted the Earth on its axis by estimates of between 10 cm (4 in) and 25 cm (10 in), increased earth's rotational speed by 1.8μs per day, and generated infrasound waves detected in perturbations of the low-orbiting GOCE satellite. Initially, the earthquake caused sinking of part of Honshu's Pacific coast by up to roughly a meter, but after about three years, the coast rose back and kept on rising to exceed the original height of the coast.

https://en.wikipedia.org/wiki/2011_T%C5%8Dhoku_earthquake_and_tsunami

Draw Illustration

Copy and Label the Illustration in the Space Provided

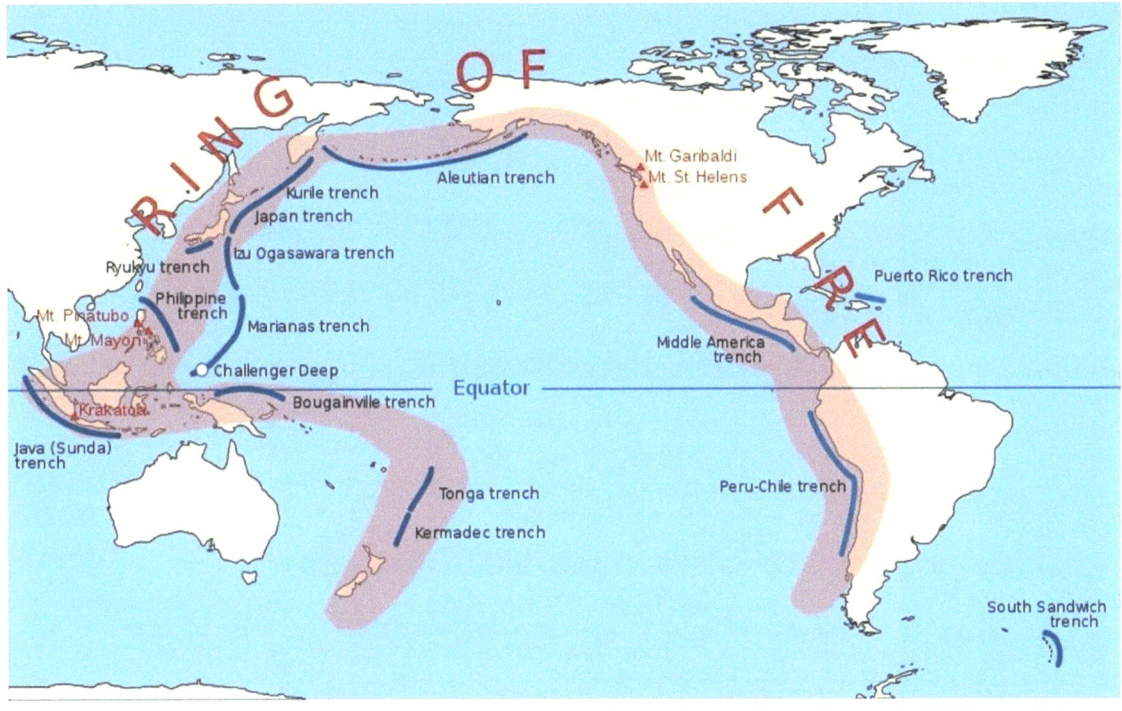

https://en.wikipedia.org/wiki/Ring_of_Fire

Draw (Copy) the Illustration Here

Interpret a Graph

Write the title of the graph _____

Circle the type of chart this represents
     Bar Chart   Line Chart   Pie Chart   Other

If applicable,
    What does the X-axis represent _____

    What does the Y-axis imply _____

Summarize what this graph represents or conveys

_____

https://pubs.usgs.gov/circ/c1187/RL

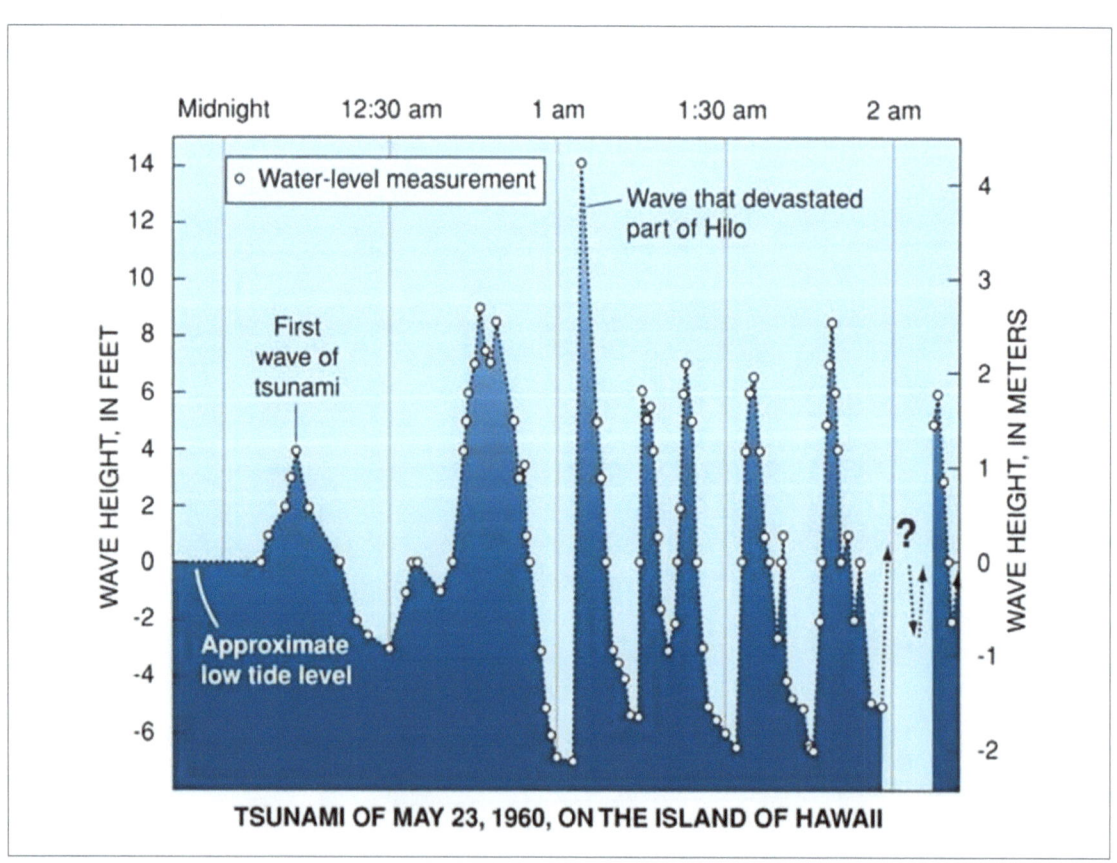

26.1 Weathering Erosion & Rivers

Summarize main points from each video.

Video Title / topic

Video Title / topic

Video Title / topic

Topic Introduction

Summarize your understanding of each paragraph.

Erosion is a process where natural forces like water, wind, ice, and gravity wear away rocks and soil. It is a geological process, and part of the rock cycle. Erosion occurs at the Earth's surface, and has no effect on the Earth's mantle and core

Most of the energy that makes erosion happen is provided by the Sun. The Sun's energy causes the movement of water and ice in the water cycle and the movement of air to create wind

Erosion can cause problems that affect humans. Soil erosion, for example, can create problems for farmers. Soil erosion can remove soil, leaving a thin layer or rocky soil behind. Erosion can also cause problems for humans by removing rocks or soil that support buildings.

Large tropical rivers like the Paraná, Indus, Brahmaputra, Ganges, Zambezi, Mississippi and the Amazon carry huge amounts of sediment down to the sea. The Nile, perhaps the world's longest river, carries much less sediment than the others.

https://simple.wikipedia.org/wiki/Erosion

Read/Summarize Text

1. **Read the passage.**
2. **Underline key expressions in each sentence.**
3. **Re-write each word (or expression) you underlined.**
4. **Summarize the passage.**

Erosion.

Ice erosion happens when a glacier moves downhill. As the ice of the glacier moves downhill, it pushes and pulls earth materials along with it. Glaciers can move very large rocks. Ice erosion can happen in another way. Cold weather causes water inside tiny cracks in rocks to freeze. As it freezes, the ice gets bigger, and pushes hard against the rock. This can break the rock. Wind erosion occurs when wind moves pieces of earth materials. Wind erosion is one of the weakest kinds of erosion. Small pieces of earth material can be rolled along the ground surface by wind. Very small pieces can be picked up and carried by the wind. Sometimes, wind can carry small pieces of earth materials over large distances. Some sediment from the Sahara Desert is carried across the Atlantic Ocean by wind.

https://simple.wikipedia.org/wiki/Erosion

Re-write words you underlined

_____   _____   _____

_____   _____   _____

Using a complete sentence, summarize or rephrase the passage

Read Text for Comprehension

Read this article for deeper understanding. No summary is required, although you may want to circle, underline, or mark key ideas and words.

Erosion mechanics

Most meanders occur in the region of a river channel with shallow gradients, a well-developed floodplain, and cohesive floodplain material. Deposition of sediment occurs on the inner edge, because the secondary flow of the river sweeps and rolls sand, rocks and other submerged objects across the bed of the river towards the inside radius of the river bend, creating a point bar below the slip-off slope. Meandering morphology is dependent upon similar bank erosion and bar growth rates.

Erosion is greater on the outside of the bend where the soil is not protected by deposits of sand and rocks. The current on the outside bend is more effective in eroding the unprotected soil, and the inside bend receives steadily increasing deposits of sand and rocks, and the meander tends to grow in the direction of the outside bend, forming a small cliff called a cut bank. This can be seen in areas where willows grow on the banks of rivers; on the inside of meanders, willows are often far from the bank, whilst on the outside of the bend, the roots of the willows are often exposed and undercut, eventually leading the trees to fall into the river. This demonstrates the river's movement. Slumping usually occurs on the concave sides of the banks resulting in mass movements such as slides.

Oxbow lakes

Oxbow lakes are created when growing meanders intersect each other and cut off a meander loop, leaving it without an active cutting stream. This process is usually linked to flooding where the river will tend to the path of least resistance. The oxbow, being of much lower energy than the more direct path, collects more and more deposited sediment each season of flooding until it becomes independent from the river. The largest oxbow lakes will be in areas with wider flood plains where the rivers have more room to meander. Over a period of time, these oxbow lakes tend to dry out or fill in with sediments.

https://en.wikipedia.org/wiki/Meander

Draw Illustration

Copy and Label the Illustration in the Space Provided

Life history of a meander

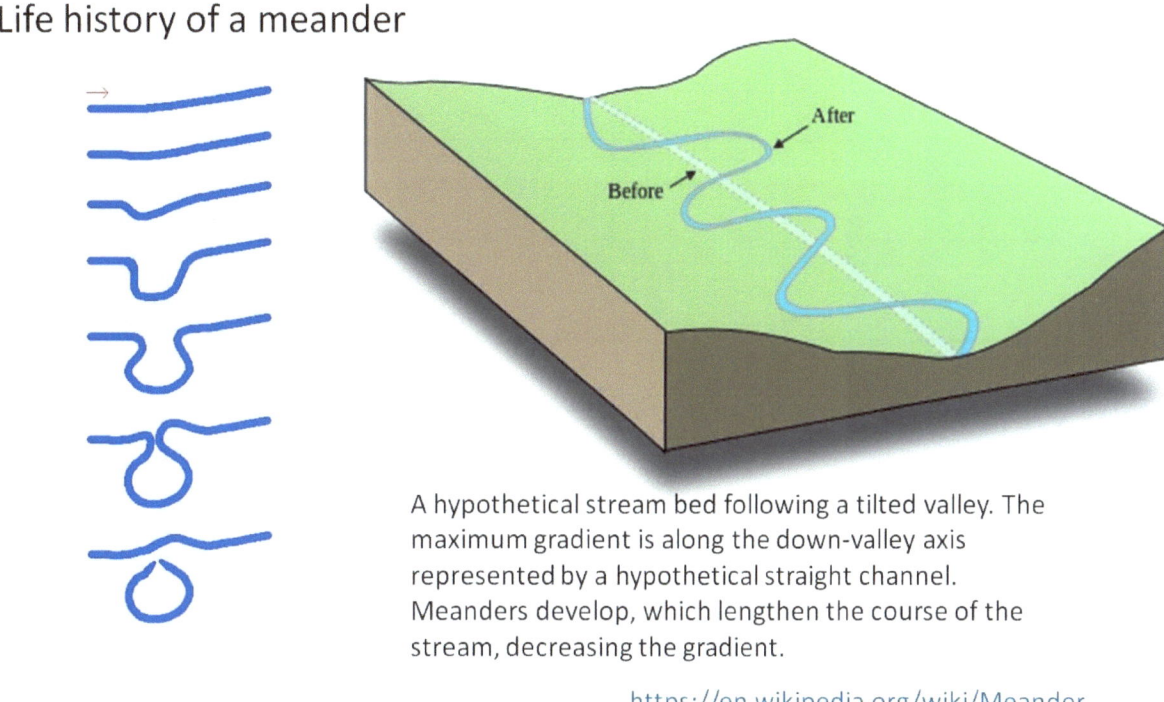

A hypothetical stream bed following a tilted valley. The maximum gradient is along the down-valley axis represented by a hypothetical straight channel. Meanders develop, which lengthen the course of the stream, decreasing the gradient.

https://en.wikipedia.org/wiki/Meander

Draw (Copy) the Illustration Here

Interpret the images

Write a 35-50 word mini-essay about the images shown below:

_____
_____
_____
_____
_____
_____

Google Images: meandering river

27.1 Agricultural Resources

Summarize main points from each video.

Video Title / topic

Video Title / topic

Video Title / topic

Topic Introduction

Summarize your understanding of each paragraph.

Agriculture is the science or practice of farming, including cultivation of the soil for the growing of crops and the rearing of animals to provide food, wool, and other products. synonyms: farming, cultivation, tillage, tilling, land/farm management, horticulture.

Agriculture is the cultivation and breeding of animals, plants and fungi for food, fiber, biofuel, medicinal plants and other products used to sustain and enhance human life.

Agriculture was the key development in the rise of sedentary human civilization, whereby farming of domesticated species created food surpluses that nurtured the development of civilization.

The major agricultural products can be broadly grouped into foods, fibers, fuels, and raw materials. Specific foods include cereals (grains), vegetables, fruits, oils, meats and spices. Fibers include cotton, wool, hemp, silk and flax. Raw materials include lumber and bamboo. ...

Read/Summarize Text

1. Read the passage.
2. Underline key expressions in each sentence.
3. Re-write each word (or expression) you underlined.
4. Summarize the passage.

Wikipedia: Agriculture

> In the past century, agriculture has been characterized by increased productivity, the substitution of synthetic fertilizers and pesticides for labor, water pollution, and farm subsidies.
>
> In recent years there has been a backlash against the external environmental effects of conventional agriculture, resulting in the organic, regenerative, and sustainable agriculture movements.
>
> The growth of organic farming has renewed research in alternative technologies such as integrated pest management and selective breeding. Recent mainstream technological developments include genetically modified food.

Re-write words you underlined

_____   _____   _____

_____   _____   _____

Using a complete sentence, summarize or rephrase the passage

Read Text for Comprehension

Read this article for deeper understanding. No summary is required, although you may want to circle, underline, or mark key ideas and words.

Livestock production systems

Animals, including horses, mules, oxen, water buffalo, camels, llamas, alpacas, donkeys, and dogs, are often used to help cultivate fields, harvest crops, wrangle other animals, and transport farm products to buyers. Animal husbandry not only refers to the breeding and raising of animals for meat or to harvest animal products (like milk, eggs, or wool) on a continual basis, but also to the breeding and care of species for work and companionship.

An ox-pulled plough in India

Livestock production systems can be defined based on feed source, as grassland-based, mixed, and landless. As of 2010, 30% of Earth's ice- and water-free area was used for producing livestock, with the sector employing approximately 1.3 billion people. Between the 1960s and the 2000s, there was a significant increase in livestock production, both by numbers and by carcass weight, especially among beef, pigs and chickens, the latter of which had production increased by almost a factor of 10. Non-meat animals, such as milk cows and egg-producing chickens, also showed significant production increases. Global cattle, sheep and goat populations are expected to continue to increase sharply through 2050. Aquaculture or fish farming, the production of fish for human consumption in confined operations, is one of the fastest growing sectors of food production, growing at an average of 9% a year between 1975 and 2007.

https://en.wikipedia.org/wiki/Agriculture

Draw Illustration

Copy and Label the Illustration in the Space Provided

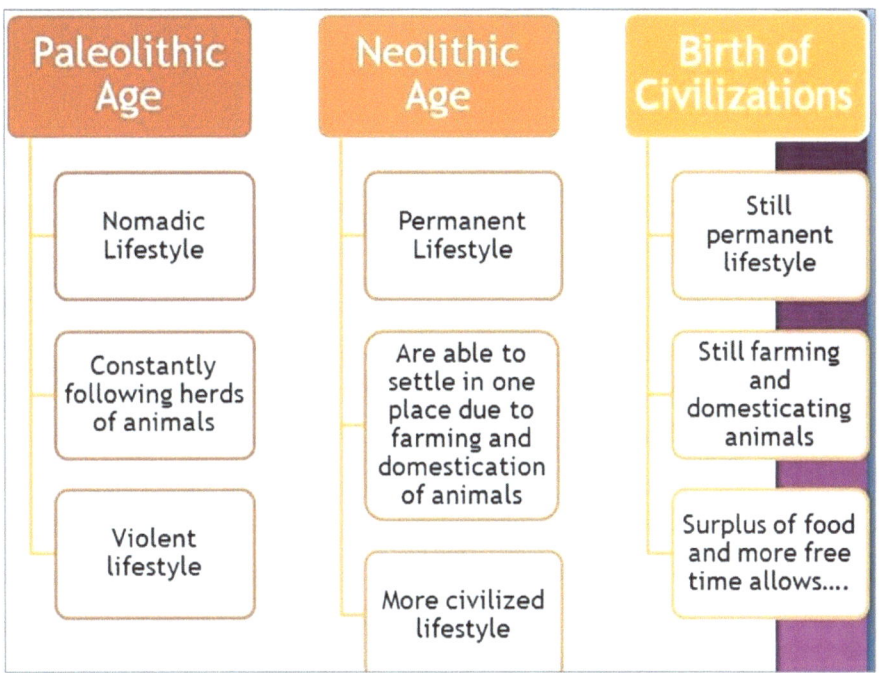

elanni.wikispaces.com

Make a Bar Graph

Top agricultural products, by crop types (million tonnes) 2004 data	
Cereals	2,263
Vegetables and melons	866
Roots and tubers	715
Milk	619
Fruit	503
Meat	259
Oilcrops	133
Fish (2001 estimate)	130
Eggs	63
Pulses	60
Vegetable fiber	30

Source:
Food and Agriculture Organization (FAO)

Using the table shown here, construct a bar graph in the space provided.

28.1 Hydrocarbons and Energy

Summarize main points from each video.

Video Title / topic

Video Title / topic

Video Title / topic

Topic Introduction

Summarize your understanding of each paragraph.

North America's massive energy diet is largely made up of hydrocarbons—a full 83 percent comes from oil, gas, and coal, and if we include nuclear energy, 91 percent comes from nonrenewable fuel sources. (as of January, 2011). http://www.resilience.org

Petroleum, natural gas, coal, renewable energy, and nuclear electric power are primary energy sources. Electricity is a secondary energy source that is generated from primary energy sources.

In 2016, energy produced in the United States was equal to about 83.9 quadrillion Btu, which was equal to about 86% of U.S. energy consumption. The difference between production and consumption was mainly in net imports of petroleum.

Natural gas production in 2016 was the second largest amount after the record high production in 2015. More efficient and cost-effective drilling and production techniques have resulted in increased production of natural gas from shale formations.

https://www.eia.gov/

Read/Summarize Text

1. **Read the passage.**
2. **Underline key expressions in each sentence.**
3. **Re-write each word (or expression) you underlined.**
4. **Summarize the passage.**

The mix of U.S. energy production changes

Total renewable energy production and consumption both reached record highs of about 10 quadrillion Btu in 2016.

Hydroelectric power production in 2016 was about 12% below the 50-year average, but increases in energy production from wind and solar helped to increase the overall energy production from renewable sources.

Energy production from wind and solar were at record highs in 2016.

https://www.eia.gov

Re-write words you underlined

_____   _____   _____

_____   _____   _____

Using a complete sentence, summarize or rephrase the passage

Read Text for Comprehension

Read this article for deeper understanding. No summary is required, although you may want to circle, underline, or mark key ideas and words.

Energy from moving air

Wind is caused by uneven heating of the earth's surface by the sun. Because the earth's surface is made up of different types of land and water, it absorbs the sun's heat at different rates. One example of this uneven heating is the daily wind cycle.

The daily wind cycle

During the day, air above the land heats up faster than air over water. Warm air over land expands and rises, and heavier, cooler air rushes in to take its place, creating wind. At night, the winds are reversed because air cools more rapidly over land than it does over water.

In the same way, the atmospheric winds that circle the earth are created because the land near the earth's equator is hotter than the land near the North Pole and the South Pole.

Wind energy for electricity generation

Today, wind energy is mainly used to generate electricity, although water pumping windmills were once used throughout the United States.

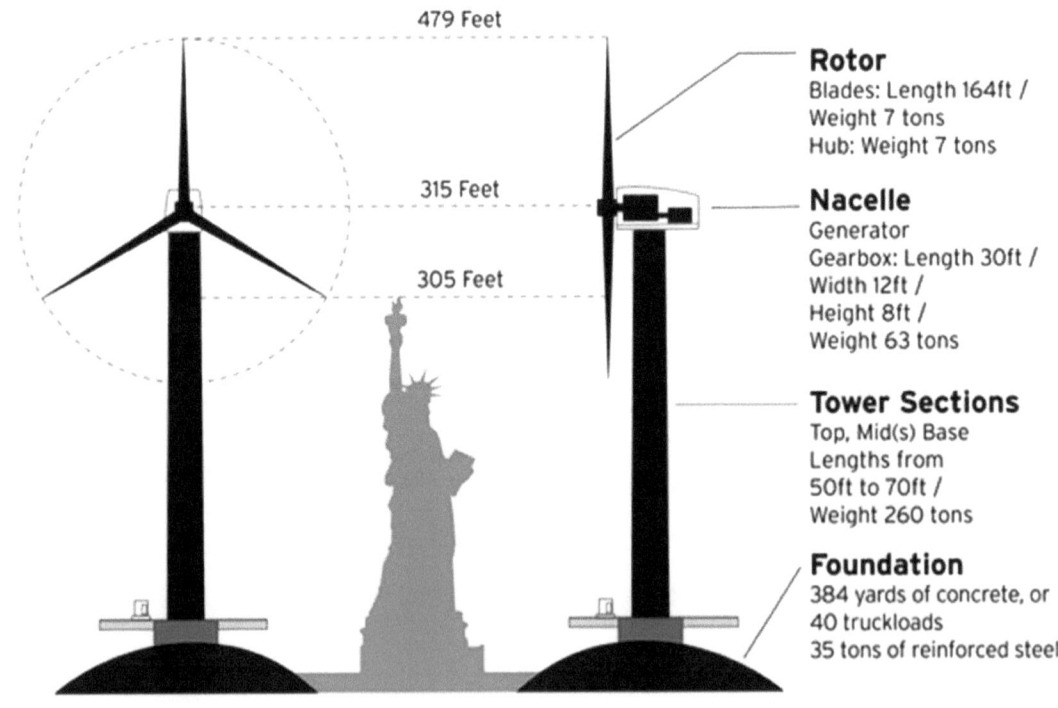

Typical wind turbine dimensions

dteenergy.com

Draw Illustration

Copy and Label the Illustration in the Space Provided

In 2016, the shares of total primary energy consumption for the five energy-consuming sectors were:

- Electric power—39%
- Transportation—29%
- Industrial—22%
- Residential—6%
- Commercial—4%

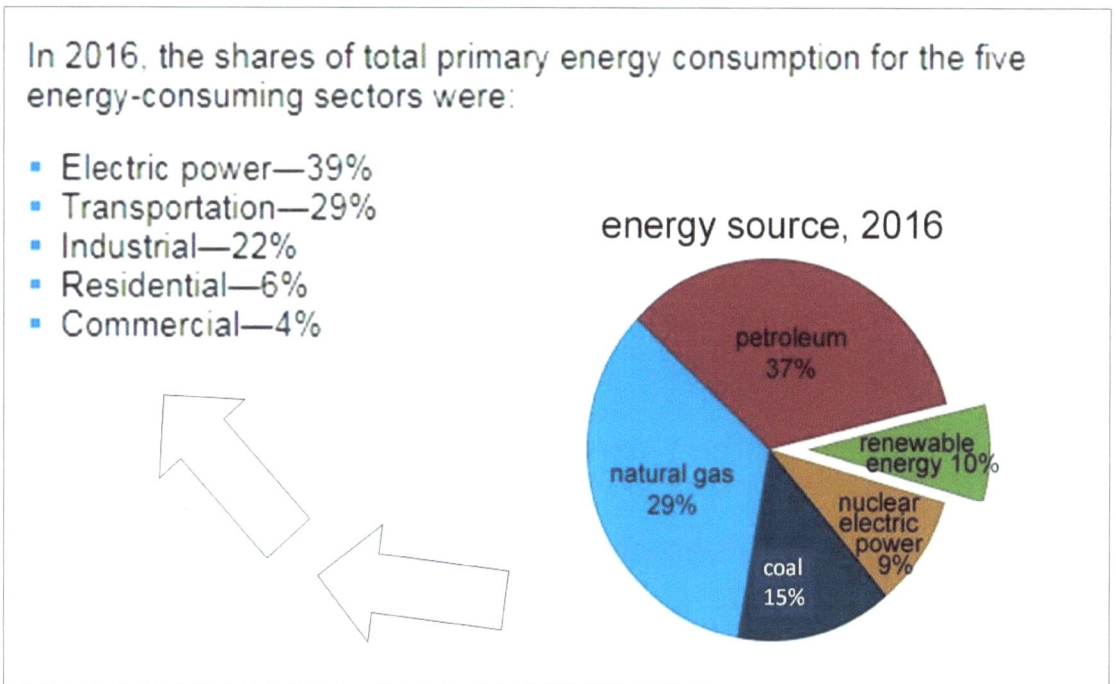

Multiple sources including D.O.E.

Draw (Copy) the Illustration Here

Interpret a Graph

Write the title of the graph _____

Circle the type of chart this represents

    Bar Chart   Line Chart   Pie Chart   Other

If applicable,

    What does the X-axis represent _____

    What does the Y-axis imply _____

Summarize what this graph represents or conveys

_____

http://www.resilience.org/wp-content/uploads/files/images/figure-17.1.PNG

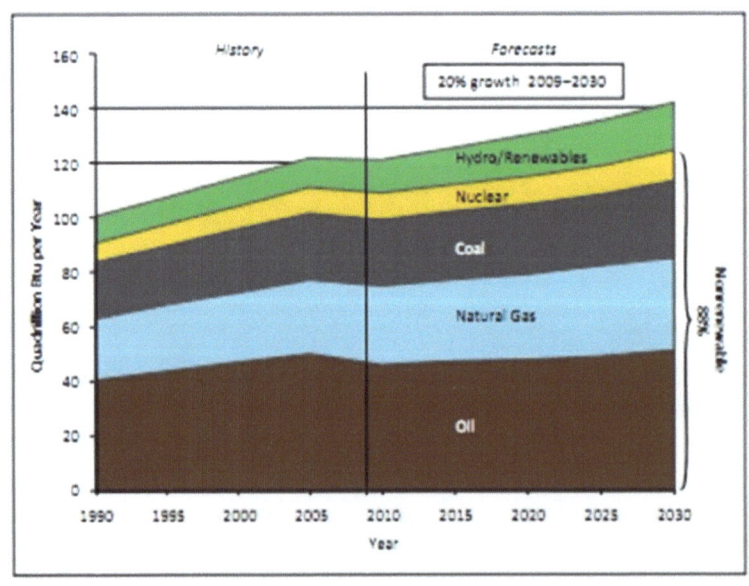

FIGURE 17.1
History and forecasts of North American energy consumption by fuel, 1990–2030.

Source: Data from U.S. Energy Information Administration, *International Energy Outlook 2009*, DOE/EIA-0484, May 27, 2009, http://www.eia.doe.gov/oiaf/ieo/.

Strategic Thinking Pages

The next four pages may be photocopied up to one time per student for each topic presented in this workbook.

Combined, they provide for strategic thinking, creativity, and collaboration/teamwork.

Show-Off Your Smarts!

Instructions

- Complete as an individual or small group.
- Discuss your ideas/answers/responses in a small group.
- Select one person to present your responses to the class.

Q1. How can this information be applied to a young-person's life?

Q2. How does this information apply to (or impact) communities?

Q3. When do scientists need to apply this information? How?

Q4. How would a person from 100 years ago view this information?

Q5. How does this topic connect to other science topics or math?

Write down at least three words introduced or covered by this topic.

1.
2.
3.
4.
5.
6.

Compare and Contrast

How would people from 1850 view this topic?
If you do not (yet) know ... use your imagination and best guess ...

Make a Poster

In the space provided here, create/draw a poster which conveys the concepts you have learned on this topic.

Write a Poem or Short Story

In the space provided write a poem or a fictional short story about the topic. The poem may rhyme ... but, it does not have to rhyme.

Add Your Notes Here ...

References

Web Sites

asherrard.us
asu.edu
bbc.co.uk
britannica.com
carleton.edu
cccblog.org
cen.acs.org
cK12.org
coolkidfacts.com
doe.gov
DOL.gov
dteenergy.com
eia.gov
geologyin.com
Hester Williams
ielts-mentor.com
learner.org
livescience.com
psu.edu
pubs.usgs.gov
decodedscience.org
elanni.wikispaces.com
geologylearn.blogspot.com
honeycuttscience.com
mayakg.blogspot.com
teachengineering.org
alternative-energy-tutorials.com
quora.com
rssweather.com
science.irank.org
sciencing.com
wikibooks.org
wikipedia.com
wikipedia.org

Text Books

Allison, M. A. (2010). Austin, TX: Holt McDougal.

Zumdahl, S. S. (2007). Belmont, CA: Brooks/Cole.

Nowiki, S. (2012). Orlando, FL: Houghton Mifflin Harcourt Publishing Company.

Dobson, K. (2008). Austin, TX: Holt, Rinehart and Winston.

For more information, visit

www.HoneycuttScience.com